LONDON MATHEMATICAL SOCIETY LECTURE NOTE SERIES

Managing Editor: Professor J.W.S. Cassels, Department of Pure Mathematics and Mathematical Statistics, University of Cambridge, 16 Mill Lane, Cambridge CB2 1SB, England

The books in the series listed below are available from booksellers, or, in case of difficulty, from Cambridge University Press.

London Mathematical Society Lecture Note Series. 134

Number Theory and Dynamical Systems

Edited by
M.M.Dodson
University of York
J.A.G.Vickers
University of Southampton

The right of the
University of Cambridge
to print and sell
all manner of books
was granted by
Henry VIII in 1534.
The University has printed
and published continuously
since 1584.

CAMBRIDGE UNIVERSITY PRESS

Cambridge

New York Port Chester Melbourne Sydney

CAMBRIDGE UNIVERSITY PRESS
Cambridge, New York, Melbourne, Madrid, Cape Town, Singapore, São Paulo

Cambridge University Press
The Edinburgh Building, Cambridge CB2 8RU, UK

Published in the United States of America by Cambridge University Press, New York

www.cambridge.org
Information on this title: www.cambridge.org/9780521369190

First published 1989
Re-issued in this digitally printed version 2008

A catalogue record for this publication is available from the British Library

ISBN 978-0-521-36919-0 paperback

Table of Contents

Contributors

J. .V. Armitage
College of St. Hild and St. Bede
University of Durham
Durham DH1 1SZ
UK

S. G. Dani
Tata Institute of Fundamental Research
Bombay 400 005
India

K. J. Falconer
School of Mathematics
University of Bristol
Bristol BS8 1TW
UK

Michel Mendes France
Mathématiques et informatique
Université de Bordeaux I
33405 Talence
France

J. Harrison
Department of Mathematics
University of California, Berkeley
Berkeley, California 94720
USA

S. J. Patterson
Mathematisches Institut
Georg-August-Universität
D-3400 Götingen
FRG

Helmut Rüssmann
Fachbereich 17 Mathematik
Johannes Gutenberg-Universität Mainz
D-6500 Mainz
FRG

S. Raghavan
Tata Institute of Fundamental Research
Bombay 400 005
India

Caroline Series
Mathematics Institute
University of Warwick
Coventry CV4 7AL
UK

J. A. G. Vickers
Department of Mathematics
University of Southhampton
Southampton SO9 5NH
UK

R. Weissauer
Fakultät für Mathematiks und Informatik
Universität Mannheim
D-6800 Mannheim
FRG

Introduction

Number theory, one of the oldest branches of mathematics, and dynamical systems, one of the newest, are rather disparate and might well be expected to have little in common. There are however many surprising connections between them. One emerged last century from the study of the stability of the solar system where the problem of "small divisors or denominators" associated with the near resonance of planetary frequencies arises and which made convergence in the series solution highly problematic. The phenomenon of "small divisors" is closely related to Diophantine approximation and it is perhaps no coincidence that Dirichlet, Kronecker and Siegel all worked on small divisor problems. These proved quite intractable until relatively recently (1942), when Siegel used ideas drawn from the theory of Diophantine approximation to overcome a problem of small divisors arising in the iteration of analytic functions near a fixed point [1]. Twenty years ago the question of the stability of the solar system was answered in more general terms by the celebrated Kolmogorov-Arnol'd-Moser theorem ([2], Appendix 8). The corresponding small divisor problem is dealt with by using Siegel's idea of imposing a suitable Diophantine inequality on the frequencies to ensure that they are not too close to resonance. Thus here Diophantine approximation again plays a central role [3].

The connection between resonance and Diophantine equations and near-resonance and Diophantine approximation is (with hindsight) a natural one. But there are other connections in quite different settings. Szemeredi's theorem that any infinite integer sequence of positive upper density contains arbitrarily long arithmetic progressions has been proved by Furstenberg using ideas from dynamical systems [4]. A conjecture of Oppenheim's on the magnitude of quadratic forms was solved recently using results about flows in hyperbolic space (see chapter 4 for further details). Some remarkable numerical relationships between eigenvalue statistics, stochastic quantum processes and zeroes of the Riemann zeta function found recently by Berry [5] are consistent with the the existence of an operator which would imply the Riemann hypothesis. The character of curves derived from incomplete θ-functions which are related to diffraction integrals and to incomplete Gauss sums depends on the arithmetic character of the exponent. When the exponent is a quadratic irrational of a certain type, the sum is fractal [6].

Another striking link between number theory and dynamical systems has been revealed recently by Vivaldi's classification of periodic orbits of generalised Arnol'd-Sinai cat maps (or hyperbolic automorphisms of the two dimensional torus) using ideals in rings of algebraic integers [7]. Extremal manifolds in the theory of non-linear Diophantine approximation play an important part in dynamical systems. In particular, Sprindžuk's solution of Mahler's problem and recent work in Diophantine approximation and Hausdorff dimension on manifolds [8] are of use in the problem of lower dimensional invariant tori (see Chapter 1) and of averaging in differential equations ([2], [9]).

These links indicated that a meeting on Number Theory and Dynamical Systems would be timely and a highly successful one was held at the University of York from March 30 to April 15 1987. The meeting consisted of a two week workshop supported by the Science and Engineering Research Council followed by a three day Conference organised with the London Mathematical Society and of more general interest. The contributions collected in this book (which does not include those of Berry [5], [6] and Vivaldi [7] mentioned above and which appear elsewhere) give some idea of the range and diversity of these connections.

The content of the KAM theorem is that perturbations of integrable non-degenerate Hamiltonian systems to a large extent retain quasi-periodic motion on invariant tori. In Chapter 1 Rüssmann considers the question of stability under general non-degeneracy conditions of the unperturbed Hamiltonian. In Chapter 2, Vickers places the iterative method, which lies at the heart of the KAM theorem and which controls the small divisors, in the natural abstract setting of a group acting on a Fréchet manifold.

Dynamical systems related to discrete groups acting on hyperbolic space can be modeled by the modular group and the associated Diophantine approximation. Patterson (Chapter 3) gives an exposition of other groups acting on hyperbolic space; these groups are arithmetic subgroups of the group of metric-preserving diffeomorphisms of the Poincaré model of hyperbolic space. Irrationals can be associated via the Markov constant with a simple loop on the punctured torus, giving a natural connection with hyperbolic geometry and with dynamics of the action of the group $PL(2, \mathbf{Z})$. This line of investigation has been carried out by Series (Chapter 4), whose paper illustrates very clearly the interplay between Diophantine approximation and hyperbolic geometry. This work is closely related to that of Patterson on Kleinian and Fuchsian groups and has some similarities with the work of Sullivan. There are also links with the work of Dani (Chapter 5) on flows in homogeneous spaces, bounded and unbounded trajectories and systems of regular and singular linear forms. Raghavan and Weissauer (Chapter 6) use estimates for Fourier coefficients of Siegel cusp forms of interest in connection with asymptotic formulae of number theoretic functions, to give estimates for the associated Satake parameters usually obtained from spherical functions.

The Hausdorff dimension gives additional information about the size of a set of Lebesgue measure 0 and is used in the investigation of exceptional sets such as well-approximable numbers and fractals. In Chapter 7, Falconer discusses the extension of Poincaré's formula from manifolds to fractal sets and makes some applications to number theoretic sets. Harrison (Chapter 8) shows that a curve associated with a quadratic irrational and constructed using an idea due to Denjoy, can be embedded in the plane and is self-similar; she also determines the Hausdorff dimension of the curve. In Chapter 9, Mendes France shows how folding paper gives rise to binary sequences which can be associated with transcendental numbers and with self-avoiding polygons. A global form of dimension for planar sets is defined which is always 2 for paper folding curves. The notions of entropy, temperature (a curve with zero temperature is a straight line segment), volume and pressure for a planar curve are introduced. An "equation of state" is presented and some novel results concerning Hausdorff dimension, entropy and chaos established.

Finally Armitage (Chapter 10) obtains a very interesting analytical connection between the Riemann zeta function and the diffusion equation and an operator arising in quantum mechanics. This approach falls short through the absence of absolute convergence of proving the Riemann Hypothesis. There is a comparison here with the intriguing behaviour mentioned above of the zeros of the zeta function.

We are very grateful to the Scientific and Engineering Research Council and to the London Mathematical Society for their generous support and encouragement, and to the University of York for the facilities and help provided. Particular thanks are also due to Ewan Kirk of Southampton for his enormous help with the typesetting of the text and to David Tranah of Cambridge University Press for his assistance and encouragement throughout the preparation of the book.

<table>
<tr><td>M. M. Dodson</td><td>J. A. G. Vickers</td></tr>
<tr><td>(York)</td><td>(Southampton)</td></tr>
</table>

References
[1] C. L. Siegel, 'Iteration of analytic functions', *Ann. Math.* **43** (1942) 607-612.
[2] V. I. Arnol'd, *Mathematical Methods of Classical Mechanics*, (Springer-Verlag, New York, 1978).
[3] M. M. Dodson and J. A. G. Vickers, 'Exceptional sets in Kolmogorov-Arnol'd-Moser theory', *J. Phys. A* **19** (1986), 349-374.
[4] H. Furstenburg, *Recurrence in Ergodic Theory and Combinatorial Number Theory*, (Princeton University Press, 1981).
[5] M. Berry, 'Riemann's zeta function: a model for quantum chaos' in *Quantum Chaos and Statistical Nuclear Physics* (Eds T. H. Seligman and H. Nishioka), Springer Lecture Notes in Physics No. 263.
[6] M. Berry and J. Goldberg, 'Renormalisation and curlicues', *Nonlinearity* **1** (1988) 1-26.
[7] F. Vivaldi, 'Arithmetic properties of strongly chaotic motions', *Physica* **25D** (1987) 105-130
[8] M. M. Dodson, B .P. Rynne and J. A. G. Vickers, 'Metric Diophantine approximation and Hausdorff dimension on manifolds', *Math. Proc. Cam. Phil. Soc.* (1989) to appear.
[9] M. M. Dodson, B .P. Rynne and J. A. G. Vickers, 'Averaging in multifrequency systems', *Nonlinearity* **2** (1989) 137-148.

1
Non-degeneracy in the perturbation theory of integrable dynamical systems

Helmut Rüssmann

Johannes Gutenberg-Universität, Mainz, FRG

§1. The Problem

The prototype of a perturbed integrable dynamical system is a Hamiltonian system

$$\frac{dx}{dt} = \frac{\partial H}{\partial y} = \left(\frac{\partial H}{\partial y_1}, \ldots, \frac{\partial H}{\partial y_n} \right),$$
$$\frac{dy}{dt} = -\frac{\partial H}{\partial x} = -\left(\frac{\partial H}{\partial x_1}, \ldots, \frac{\partial H}{\partial x_n} \right) \tag{1}$$

where

$$x = (x_1, \ldots, x_n) \in T^n = \mathbf{R}^n / 2\pi \mathbf{Z}^n$$

is the vector of angular variables and

$$y = (y_1, \ldots, y_n) \in B \subseteq \mathbf{R}^n$$

is the vector of action variables varying in some open set B of $\mathbf{R}^n$. The Hamiltonian

$$H : T^n \times B \to \mathbf{R}$$

is supposed to be real analytic and of the form

$$H(x, y) = H_0(y) + H_1(x, y) \tag{2}$$

where

$$|H_1(x, y)| \leq \epsilon \quad \text{for all } x \in T^n, \ y \in B \tag{3}$$

with $\epsilon > 0$ sufficiently small.

The unperturbed system ($\epsilon = 0$)

$$\frac{dx}{dt} = \omega(y) := \frac{\partial H_0}{\partial y}(y), \qquad \frac{dy}{dt} = 0$$

has nothing but quasiperiodic phase trajectories

$$x = t\omega(b) + \text{const} \tag{4}$$

lying on invariant tori

$$y = b \in B. \tag{5}$$

In the case that the frequency vector $\omega(b) = (\omega_1(b), \ldots, \omega_n(b))$ is non-resonant, that is

$$< k_1, \omega(b) > \; = k_1\omega_1(b) + \ldots + k_n\omega_n(b) \neq 0 \quad \text{for all}$$

$$k = (k_1, \ldots, k_n) \in \mathsf{N}_0^n \setminus \{0\}, \quad \mathsf{N}_0 = \{0, 1, 2, \ldots\},$$

the quasiperiodic phase trajectories (4) fill densely the invariant torus $T^n \times \{b\}$, on which they lie.

The question is under which conditions these invariant tori survive in the perturbed system ($0 < \epsilon \ll 1$). The survival can roughly be described as follows:

The torus $T^n \times \{b\}$ is slightly deformed,

$$y = b + U(x, b) \tag{6}$$

where

$$U : T^n \times B \to \mathsf{R}^n$$

is continuous and we have

$$\sup_{x \in T^n} |U(x, b)| \longrightarrow 0 \quad \text{for} \quad \epsilon \longrightarrow 0 \quad \text{and} \quad b \in B. \tag{7}$$

Moreover $U = (U_1, \ldots, U_n)$ has partial derivatives with respect to $x_1, \ldots, x_n$, and we have

$$\frac{\partial U_j}{\partial x_l} = \frac{\partial U_l}{\partial x_j} \quad .$$

The frequency vector is also slightly deformed, but is non-resonant, so that the phase trajectories

$$x = t\bar{\omega}(b) + \text{const} \quad t \in \mathsf{R} \tag{8}$$

fill (6) densely, and $\overline{\omega} : B \to \mathbf{R}^n$ is supposed to be at least of class C^1 and to satisfy

$$|\omega(b) - \overline{\omega}(b)| \longrightarrow 0 \quad \text{for} \quad \epsilon \longrightarrow 0. \tag{9}$$

The well known theorem of A.N.Kolmogorov [4, p.183] states that a sufficient condition for the survival of most of the invariant tori (5) in the form (6), (7) is

$$\det \frac{\partial \omega}{\partial y}(b) = \det(\frac{\partial^2 H_0}{\partial y_j \partial y_l}(b)) \neq 0 \quad \text{for all} \quad b \in B. \tag{10}$$

As Moser [8, p.44] has shown this theorem is true even with $\overline{\omega} = \omega$.

The goal of this paper is to formulate infinitely many further sufficient conditions which all together lead to a very natural non-degeneracy condition for ω, respectively H_0, guaranteeing the existence of invariant tori of the perturbed system (1),(2),(3), $0 < \epsilon \ll 1$.

§2. A general non-degeneracy condition

In the following definition the frequency vector ω is not necessarily the gradient of a Hamiltonian H_0 as above.

Definition 1. *Let B be an open and connected subset of $\mathbf{R}^n$ and $\omega : B \to \mathbf{R}^n$ be a real analytic vector function. We call ω non-degenerate if the range $\omega(B)$ of ω does not lie in an $(n-1)$-dimensional linear subspace of $\mathbf{R}^n$. If $\omega(B)$ belongs to such a subspace ω is called degenerate.*

Remark: In the case $n = 1$, a real analytic function $\omega : B \to \mathbf{R}$ with B an open interval in $\mathbf{R}$ is non-degenerate if and only if it does not vanish identically in B. Now from function theory it is well known that if ω vanishes in B then at every point $b \in B$ all coefficients of the Taylor expansion of ω at b vanish and conversely, if all coefficients of the Taylor expansion of ω vanish at some point $b \in B$ then ω vanishes identically in B. So for non-degeneracy of ω in the case $n = 1$ it is sufficient that the Taylor expansion of ω at some point $b \in B$ contains at least one non-vanishing coefficient, and necessary that for each point $b \in B$ the Taylor expansion of ω in b contains at least one non-vanishing coefficient.

The corresponding facts for arbitrary n are formulated in the following:

Lemma 1. *Let $B \subseteq \mathbf{R}^n$ be open and connected. Then for a real analytic function $\omega : B \to \mathbf{R}^n$ to be non-degenerate it is sufficient that the Taylor expansion*

$$\omega(y) = \sum_{l \in \mathbf{N}_0^n} \frac{(y-b)^l}{l!} \omega^{(l)}(b) \tag{11}$$

of ω at some point $b \in B$ contains n linearly independent coefficients

$$\omega^{(l)}(b) \in \mathbf{R}, \qquad l = l^{(1)}, \dots, l^{(n)} \in \mathbf{N}_0^n, \tag{12}$$

and it is necessary that the Taylor expansion (11) contains n linearly independent coefficients (12) at each point $b \in B$ where the choice of the indices $l^{(\alpha)} = l^{(\alpha)}(b)$, $\alpha = 1, \dots, n$ may depend on $b \in B$.

Proof: We prove the corresponding statement for degeneracy. We fix b and assume that the series (11) does not contain n linearly independent coefficients (12). Then the linear space V spanned by all Taylor coefficients $\omega^{(l)}(b)$, $l \in \mathbf{N}_0^n$ has dimension $\leq n-1$. We consider $\mathbf{R}^n$ as a Euclidean space provided with usual inner product $< ., . >$ and choose an orthonormal base $\{e_1, \dots, e_n\}$ such that V is contained in the linear subspace spanned by $e_1, \dots, e_{n-1}$. Then we can write

$$\omega^{(l)}(b) = \sum_{j=1}^{n-1} a_{jl} e_j \ , \qquad l \in \mathbf{N}_0^n \tag{13}$$

with uniquely determined $a_{jl} \in \mathbf{R}$. Inserting (13) in (11) we get

$$\omega(y) = \sum_{j=1}^{n-1} \omega_j(y) e_j \tag{14}$$

with

$$\omega_j(y) = \sum_{l \in \mathbf{N}_0^n} \frac{(y-b)^l}{l!} a_{jl} \ , \qquad j = 1, \dots, n-1.$$

From (13) we obtain the estimate

$$|\omega^{(l)}(b)| = \sqrt{\sum_{j=1}^{n-1} a_{jl}^2} \ \geq |a_{jl}|$$

which makes clear that each series $\omega_j(y)$ is absolutely convergent in every polydisc

$$|y - b| = \max_{1 \leq j \leq n} |y_j - b_j| \leq r$$

in which the series (11) converges absolutely. So $\omega_1, \ldots, \omega_{n-1}$ are analytic in a certain open neighbourhood W of b.

Now ω is analytic in B. Therefore the functions

$$y \mapsto \omega_j(y) = \,< \omega(y), e_j >, \qquad j = 1, \ldots, n$$

are analytic in B. Comparing

$$\omega(y) = \sum_{j=1}^{n} \omega_j(y) e_j$$

with (14) we get $\omega_n(y) = 0$ in the neighbourhood W of b. Since B is connected we have $\omega_n(y) = 0$ for all $y \in B$. As a consequence $\omega(B)$ lies in an $n - 1$-dimensional linear subspace of $\mathbf{R}^n$. Conversely, if this is the case we choose an orthonormal base $\{e_1, \ldots, e_{n-1}\}$ of this subspace. So we obtain (14) with coefficients

$$y \mapsto \omega_j(y) = \,< \omega(y), e_j >, \qquad j = 1, \ldots, n-1$$

analytic in B. Now we differentiate (14) at an arbitrary point $b \in B$ in order to obtain (13) with

$$a_{jl} = \omega_j^{(l)}(b).$$

But (13) means that all Taylor coefficients at b belong to an $(n - 1)$-dimensional linear subspace of $\mathbf{R}^n$. So n linearly independent coefficients (12) cannot be found, and the lemma is proved.

Definition 2. *Let B be an open and connected subset of $\mathbf{R}^n$ and $H_0 : B \to \mathbf{R}$ be a real analytic function. We call H_0 non-degenerate if the gradient*

$$y \mapsto \omega(y) = \frac{\partial H_0}{\partial y}(y) : B \to \mathbf{R}^n$$

of H_0 is non-degenerate; otherwise H_0 is called degenerate.

Lemma 2. *Let $B \subseteq \mathbf{R}^n$ be open and connected. A real analytic function $H_0 : B \to \mathbf{R}$ is non-degenerate if and only if it does not depend - up to a linear change of variables - on less than n variables.*

Proof: H_0 is degenerate if and only if the range of $\omega = \frac{\partial H_0}{\partial y}$ lies in an $(n-1)$-dimensional subspace V of $\mathbf{R}^n$. We assume that this is the case and choose an orthonormal base $\{e_1, \ldots, e_n\}$ of $\mathbf{R}^n$ such that V lies in the span of $e_1, \ldots, e_{n-1}$. Then we have

$$< \omega(y), e_n > \; = 0 \quad \text{for all } y \in B, \tag{15}$$

hence, putting

$$y = zC = z_1 e_1 + \ldots + z_n e_n, \tag{16}$$

$$\frac{\partial H_0(zC)}{\partial z_n} = \; < \frac{\partial H_0}{\partial y}(zC), e_n > \; = 0 \quad \text{for all } z \in BC^{-1}. \tag{17}$$

So the function

$$z = (z_1, \ldots, z_n) \mapsto H_0(zC)$$

does not depend on z_n in its domain of definition BC^{-1}. Conversely if we have (17) for a linear transformation (16) where only

$$\det(e_1, \ldots, e_n) \neq 0$$

is required, we get (15), and consequently $\omega(B)$ lies in an $(n-1)$-dimensional subspace of $\mathbf{R}^n$. The lemma is proved.

Examples: We put $B = \mathbf{R}^n$ and define H_0 by

$$(1) \qquad H_0(y) = y_1 y_1 + y_2 y_1^2 + y_3 y_1^3 + \ldots + y_n y_1^n.$$

Then H_0 is non-degenerate by Lemma 1 because we have

$$\omega(y) = \frac{\partial H_0}{\partial y}(y) = (2y_1, y_1^2, \ldots, y_1^n),$$

and so

$$\det(\frac{\partial \omega}{\partial y_1}, \frac{\partial^2 \omega}{\partial y_1{}^2}, \ldots, \frac{\partial^n \omega}{\partial y_1{}^n})$$

$$= \begin{vmatrix} 2 & \cdot & \cdot & \cdots & \cdot \\ 0 & 2! & \cdot & \cdots & \cdot \\ 0 & 0 & 3! & \cdots & \cdot \\ \vdots & \vdots & \vdots & \ddots & \vdots \\ 0 & 0 & 0 & \cdots & n! \end{vmatrix} = 2(2!)\ldots(n!) \neq 0.$$

$$(2) \qquad H_0(y) = (y_1 + \ldots + y_n)y_1^2.$$

Here H_0 is degenerate for $n > 2$ by Lemma 2. For $n = 2$ H_0 is non-degenerate by Lemma 1 because, for example,

$$\det(\omega, \frac{\partial \omega}{\partial y_2})(1,1) = \begin{vmatrix} 5 & 2 \\ 1 & 0 \end{vmatrix} = -2.$$

§3. Formulation of the existence theorem

We have the following

Theorem. *Let B be an open and connected subset of $\mathbf{R}^n$ and let the unperturbed Hamiltonian $H_0 : B \to \mathbf{R}$ of system (1),(2),(3) be non-degenerate. Then most of the invariant tori $T^n \times \{b\}$, $b \in B$ of the unperturbed system ($\epsilon = 0$) survive in the form (6),(7). These surviving tori form a majority in the sense that given any open set K with compact closure $\overline{K} \subseteq B$ then the Lebesgue measure of the set*

$K_\epsilon = \{b \in K : \text{there is an}$
$\qquad\qquad\qquad \text{invariant torus (6) of (1), (2), (3) with (7)}\}$

tends to the full measure of K:

$$\mu(K_\epsilon) \nearrow \mu(K) \quad \text{as} \quad \epsilon \longrightarrow 0.$$

Moreover the frequency vector $\overline{\omega}$ of the flow (8) on the invariant torus (6), $b \in K_\epsilon$ satisfies the relations (9) and

$$|<k, \overline{\omega}(b)>| \geq c(\epsilon)e^{-\sqrt{|k|}} \quad \text{for all } k \in \mathbf{Z}^n \setminus \{0\} \qquad (18)$$

where $0 < c(\epsilon) \longrightarrow 0$ *as* $\epsilon \longrightarrow 0$.

A more quantitative formulation of this theorem and a complete proof of it will appear as part of our work on small divisors in which the Newton method is avoided. Indeed, the classical technique of iteration of celestial mechanics is used. But instead of linearisation at each step of the iteration process we construct a fixed point of a certain functional operator which simultaneously gives a generating function as well as the canonical transformation belonging to it. The advantage of this method for small divisor problems can be compared with Picard-Lindelöf iteration as the dominating method for proving the existence of solutions of ordinary differential equations whereas the Newton method is used mostly for numerical purposes.

It has often been said that the rapid convergence of the Newton iteration is necessary for compensating the influence of small divisors like those in (18). But a deeper analysis shows that this is not true. Also the results of Aubry [1], Herman [5], and Mather [7] (see also Katok [6]) on special classes of small divisor problems point in this direction.

Historically the Newton method was surely necessary to establish the main theorems of the KAM-theory. But for clarifying the structure of small divisor problems the Newton method is not useful because it compensates not only the influence of small divisors but also many bad estimates veiling the true structure of the problems.

Our method leads, in all analytic soluble small divisor problems, to the natural bound

$$\int_1^\infty \log \frac{1}{M_s} \frac{ds}{s^2} < \infty \qquad (19)$$

where

$$M_s = \min_{0 < |k| \leq s} | < k, \omega > | \quad > 0 \qquad (20)$$

and ω is the frequency vector of the problem. Also in (18) the function $s \mapsto \exp(-s^{1/2})$ can be replaced by a monotonically decreasing function $s \mapsto M_s$ satisfying (19).

For problems near a singular point condition (19) is equivalent to Bryuno's condition [2]. So we cannot obtain more than Bryuno got by means of the Newton method besides more clarity in the formulae. However, in some cases we are able to make the length of the iteration steps arbitrarily small such that in the limit we integrate in the presence of small divisors in the sense that the iteration process can be considered as a Riemannian sum which tends to a Riemann integral, provided (19) is satisfied.

In problems on the torus the situation is much more difficult. But using the theory of elliptic functions we are able to show that in the presence of small divisors of constant type, that is we have $sM_s \geq \text{const} > 0$ for $s > 0$ in (20), the length of the iteration steps may also tend to zero, in some cases and this relates our work to the work of M.R.Herman [5].

§4. Lower dimensional invariant tori

We still consider the more general Hamiltonian system

$$\frac{dx}{dt} = \frac{\partial H}{\partial y}, \qquad \frac{du}{dt} = \frac{\partial H}{\partial v},$$
$$\frac{dy}{dt} = -\frac{\partial H}{\partial x}, \qquad \frac{dv}{dt} = -\frac{\partial H}{\partial u}, \qquad (21)$$

where

$$H : T^n \times B \times W \to \mathbf{R}$$

is real analytic, $B \subseteq \mathbf{R}^n$ is open and connected, and

$$W = \{w = (u,v) \in \mathbf{R}^m \times \mathbf{R}^m : |w| < r\}$$

is some neighbourhood of $w = 0$ in $\mathbf{R}^m \times \mathbf{R}^m$. Moreover H has the form

$$H = H_0(y) + \frac{1}{2} < w, wP(y) > + H_2(x,y,w) \qquad (22)$$

where $< .,. >$ is the usual inner product of $\mathbf{R}^m \times \mathbf{R}^m$ and $P(y)$ is a symmetric $2m \times 2m$-matrix analytically depending on y in B. We assume

$$|H_2(x,y,w)| \leq \epsilon \quad \text{for all } x \in T^n, \quad y \in B, \quad w \in W. \qquad (23)$$

The unperturbed system ($\epsilon = 0$) has the invariant tori

$$y = b \in B, \ w = 0$$

with the quasiperiodic trajectories (4). Also in this extended case the question of survival to invariant tori of (21),(22),(23) for small $\epsilon > 0$ arises.

An unavoidable condition for the survival seems to be that the purely imaginary eigenvalues

$$\pm i\Omega_j, \quad \Omega_j : B \to \mathbf{R}, \quad j = 1,\ldots,m_0 \le m$$

of the w-part of the unperturbed system ($\epsilon = 0$)

$$\frac{dx}{dt} = \frac{\partial H_0}{\partial y}(y), \quad \frac{dy}{dt} = 0,$$

$$\frac{dw}{dt} = wP(y)J, \quad J = \begin{pmatrix} 0 & -I \\ I & 0 \end{pmatrix} \tag{24}$$

are different from zero and different from one another:

$$(\Omega_j - \Omega_l)(y) \ne 0, \quad y \in B, \quad j \ne l,$$
$$\Omega_j(y) \ne 0; \quad j,l = 1,\ldots,m_0. \tag{25}$$

There have appeared papers of Eliasson [3] and Pöschel [9] in which the existence of invariant tori for the perturbed system (21),(22),(23) for small $\epsilon > 0$ are constructed in the case $m_0 = m$, that is, only purely imaginary eigenvalues in (24) exist. Moreover it is assumed that (10) is valid and then some other conditions are required which we do not specify here because we like to remark that all such additional conditions are superfluous even in the general case $1 \le m_0 < m$ and that H_0 has only to be non-degenerate.

The reason for this is that besides (18) the Diophantine inequalities

$$|< k,\overline{\omega}(b) > + \overline{\Omega}(b)| \ge c(\epsilon)e^{-\sqrt{|k|}} \quad \text{for all } k \in \mathbf{Z}^n \tag{26}$$

have to be satisfied for sufficiently many $b \in B$ and for $(\overline{\omega}, \overline{\Omega})$ sufficiently near by (ω, Ω) where $\Omega : B \to \mathbf{R}$ is a real analytic function standing for Ω_j or $\Omega_j - \Omega_l$ in (25). Now (26) can be satisfied in the same way as (18) provided that ω/Ω is non-degenerate, and this is the case if and only if ω is non-degenerate because of (25). So the non-degeneracy of H_0 is the only condition guaranteeing the existence of invariant tori of the perturbed system (21),(22),(23) for small $\epsilon > 0$.

The main problem in the proof of the theorem above and its extension to lower dimensional invariant tori is the "approximation of dependent quantities" as Sprindžuk [11] calls problems of Diophantine approximation with too few unknowns. In the theory of dynamical systems there are two problems of this sort: the degeneracy problem for the angular variables, that is, there are not enough free parameters for controlling the frequencies, and the existence of lower dimensional tori, and both problems are transcendentally connected in the analytic case, as we have indicated above.

The first problem is much more difficult than the second one because the implicit function theorem which is the basic tool in KAM-theory to control the frequencies has to be avoided.

In our talks in Moser's seminar in Zürich (January 1986) and at the Number Theory and Dynamical Systems conference in York (April 1987) we mainly treated the second problem because we could still not deal with the first one. In the course of the year 1987 we proved the theorem in section 3 for a class of n-tuplets of independent Taylor coefficients (12) and we finally succeeded in our talk at the IVth German-French Meeting on Mathematical Physics in Marseille (March 1988) in proving the existence theorem under the condition of mere non-degeneracy of the unperturbed Hamiltonian.

The extension of this theorem to the problem of lower dimensional tori is straight forward provided only purely imaginary eigenvalues are present. In the general case new difficulties arise because the semigroup approach frequently used in hyperbolic problems is not applicable in the presence of complex frequencies. The same difficulties arise if one likes to construct the stable and unstable manifold in dissipative systems not by means of real analysis according to Hadamard and Perron but by complex analysis.

In order to handle such problems we studied intensively linear equations containing small divisors produced by complex frequencies. It turned out that better results can be obtained by means of the best approximation of differentiable almost-periodic functions in two variables by trigonometrical polynomials than by using the geometry of numbers to count lattice points. We developed such a theory of best approximation in two variables, not available in the literature, and used it in an identity for the Laplacian in the plane which is verified by evaluating iterated integrals. See our contribution to the volume on the occasion of Moser's 60th birthday.

§5. The twist mapping theorem

Our non-degeneracy conditions for Hamiltonian systems are completely analogous to those sufficient for the existence of invariant curves of area preserving mappings of an annulus. We consider a perturbed twist mapping

$$M : \begin{cases} x_1 = x + h(y) + f(x,y) \\ y_1 = y \qquad\quad + g(x,y) \end{cases}$$

where

$$f, g : T^1 \times B \to \mathbf{R}, \quad B = (\alpha, \beta), \quad 0 < \alpha < \beta$$

are real analytic functions satisfying the estimate

$$|f(x,y)| + |g(x,y)| \le \epsilon, \quad \text{for all } (x,y) \in T^1 \times B. \tag{27}$$

The function $h : B \to \mathbf{R}$ is assumed to be real analytic and to satisfy the condition

$$\frac{dh}{dy}(y) \ne 0, \quad \text{for all } y \in B. \tag{28}$$

The unperturbed mapping ($\epsilon = 0$) has invariant circles $y_1 = y = b \in B$.

Moser [8] has proved that many of these invariant circles survive as invariant curves of the mapping M with $\epsilon > 0$ sufficiently small in (27) provided M is not necessarily area preserving but has the intersection property. This means that simply closed curves in $T^1 \times B$ near circles $T^1 \times \{b\}$, $b \in B$ intersect its image under M.

The condition (28) corresponds to condition (10) for Hamiltonian systems. Non-degeneracy of h in the broadest sense corresponding to our considerations above means here: dh/dy must not vanish identically in B, that is locally, at each point $b \in B$ the Taylor expansion

$$h(y) = \sum_{j=0}^{\infty} \frac{h^{(j)}(b)}{j!}(y - b)^j, \quad h^{(j)}(b) \neq 0, \quad j = j(b) \geq 1$$

has at least one non-vanishing coefficient $h^{(j)}(b) \neq 0$ for some $j = j(b) \geq 1$. This situation can easily be reduced to the case (28) as Moser has shown: one introduces a new radial variable

$$z = h(y),$$

and this equation can be locally inverted. The intersection property remains valid under such transformations.

In general the twist mapping theorem has been formulated for the existence of individual invariant curves. But it is not difficult, to give also a measure theoretic version similar to the theorem in section 3 using the same tools as for the individual case.

We conclude with the remark that similar to the considerations of this paper the condition of non-degeneracy can be weakened in the case of invariant tori lying on an energy surface and in the case of Arnol'd's theorem [4, p.185] for applications in celestial mechanics.

References

[1] S. Aubrey and P. Y. Le Daeron, 'The discrete Frenkel-Kontorova model and its extensions I', *Physica* **8D** (1983) 381–422.

[2] A. D. Bryuno, 'Analytic Form of Differential Equations', *Trudy MMO* **25** (1971) 119-262 *Translations of the Moscow Math. Soc.* **25** (1971) 131–288.

[3] L. Eliasson, 'Perturbations of Stable Invariant Tori', Preprint (1985).

[4] *Encyclopaedia of Mathematical Sciences, Vol. III*,(Springer-Verlag, 1988).

[5] M. R. Herman, 'Sur les Courbes Invariantes par les Difféomorphismes de l'anneau I', *Astérisque*,**103-104** .

[6] A. Katok, 'Some remarks on Birkhoff and Mather twist map theorems', *Ergodic Theory and Dynamical Systems* **2** (1982) 185–194.

[7] J. N. Mather, 'Existence of quasi-periodic orbits for twist homeomorphisms of the annulus', *Topology* **21** (1982) 457–467.

[8] J. Moser, 'Stable and Random Motions in Dynamical Systems', *Annals of Mathematical Studies* **77** Princeton University Press, 1973.

[9] J. Pöschel, 'On Elliptic Lower Dimensional Tori in Hamiltonian Systems', Preprint (1988), Universität Bonn.

[10] C. L. Siegel and J. K. Moser, *Lectures on Celestial Mechanics* (Springer-Verlag, 1971).

[11] V. G. Sprindžuk, *Metric Theory of Diophantine Approximations* (John Wiley & Sons, 1979).

2
Infinite dimensional inverse function theorems and small divisors

J.A.G.Vickers

University of Southampton, Southampton, UK

§1. Introduction.

In the study of dynamics one is very often interested in the stability of fixed points or invariant sets. For example the system of differential eqations

$$\frac{dz}{dt} = f(z) \tag{1.1}$$

where $z = (z_1, \ldots, z_n) \in \mathbf{C}^n$ and $f : \mathbf{C}^n \to \mathbf{C}^n$ is an analytic function with $f(0) = 0$, has $z(t) = 0$ as a solution. This solution is said to be past and future stable (henceforth just called stable) if points near the origin remain near the origin under evolution by (1.1) to both the past and future. More precisely we say $z = 0$ is stable if for every neighbourhood U of 0 there exists a neighbourhood V with $0 \in V \subset U$ such that $z(0) \in V$ implies $z(t) \in U$ for all $t \in \mathbf{R}$.

It follows from a well known theorem due to Liapunov [1] that a necessary condition for the future stability of $z = 0$ is that the eigenvalues of Df_0 have non-positive real part, while for past stability they must have non-negative real part. Thus for $z = 0$ to be stable the eigenvalues must be purely imaginary. In fact a theorem of Carathéodory and Cartan gives necessary and sufficient conditions for the stability of $z = 0$ [2].

Theorem 1.1. *The solution $z = 0$ of the differential equation (1.1) is stable if and only if there exists a holomorphic transformation $g : \mathbf{C}^n \to \mathbf{C}^n$, $z \mapsto \zeta$ with $g(0) = 0$ and $Dg_0 = I$ which brings (1.1) to the form*

$$\frac{d\zeta}{dt} = A\zeta \tag{1.2}$$

where A is the Jacobian of f at 0 and in addition A is diagonalisable with purely imaginary eigenvalues.

We now give a slightly different interpretation to this result by regarding (1.1) as giving the integral curves of the holomorphic vector field

$$X = \sum_{k=1}^{n} f_k(z)\frac{\partial}{\partial z_k}. \tag{1.3}$$

Let $\mathcal{M}$ be the set of such holomorphic vector fields which vanish at the origin, and let $\mathcal{G}$ be the group of holomorphic transformations which vanish at the origin and have $Dg_0 = I$. Then there is a natural action $\mathcal{G} \times \mathcal{M} \to \mathcal{M}$ of $\mathcal{G}$ on $\mathcal{M}$ given by

$$(g, X) \mapsto g.X$$

where $g.X(z) = g_*(X(g^{-1}(z)))$

and g_* is the induced map on the tangent bundle.

We say two vector fields X and Y are (holomorphically) equivalent if

$$X = g.Y \quad \text{for some } g \in \mathcal{G}.$$

Given a vector field $X \in \mathcal{M}$, the linear part $(X)_1$ is given by

$$(X)_1 = \sum_{j,k=1}^{n} a_{jk} z_j \frac{\partial}{\partial z_k} \tag{1.4}$$

where

$$a_{jk} = \frac{\partial f_k}{\partial z_j}(0).$$

In this language the condition that the differential equation (1.1) can be brought to the form (1.2) is just the condition that there exists a g such that

$$Ag(z) = A\zeta = \frac{d\zeta}{dt} = \frac{dg(z)}{dt} = g_*(\frac{dz}{dt}) = g_* f(z)$$

i.e., such that

$$A\zeta = g_* f(g^{-1}(\zeta))$$

which is just the condition that the corresponding vector field X is equivalent to its linear part $(X)_1$. Thus the stability of the differential equation (1.1) is intimately connected with the theory of normal forms of holomorphic vctor fields and in particular with the structure of the orbits of the linear vector fields under the action of $\mathcal{G}$. It is very difficult to give necessary and sufficient conditions which determine when a vector field is equivalent to its linear part. Indeed the conditions required are the most delicate in the situation in which we are most interested, when the eigenvalues are purely imaginary, but one can at least find a set of sufficient conditions which guarantee that the vector field is equivalent to its linear part and hence the corresponding differential equation is stable.

There are many other situations in dynamics where a problem in stability is closely related to a problem of normal forms. For example the stability of a fixed point of an analytic homeomorphism under iteration (to the past and future) is equivalent to the biholomorphic equivalence of the mapping to a diagonalisable linear transformation with eigenvalue of modulus one. See [3] for an extensive rewiew of this problem. A second important example is the KAM theorem which gives conditions under which invariant tori persist when an integrable Hamiltonian system is perturbed and is very closely related to the problem of the normal forms for the associated Hamiltonian vector fields under the action of the group of symplectic diffeomorphisms.

Rather than investigate the orbits of points of $\mathcal{M}$ under the action of $\mathcal{G}$ directly, it is much easier to look at the induced linear map between tangent spaces and deduce the local structure of the orbits using the inverse function theorem. Unfortunately neither $\mathcal{M}$ nor $\mathcal{G}$ have the structure of Banach spaces and in general one cannot appy the inverse function theorem without imposing some additional conditions. It turns out that each orbit can be associated with with a point in a *finite* dimensional Euclidean space and one can apply the inverse function theorem provided this point satisfies an appropriate Diophantine approximation property. This is closely related to the existence or otherwise of "small divisors" in the series solution to the the problem. A feature of this approach is the way that the arithmetic forms associated with the Diophantine condition arise naturally from the eigenvalues of the linearised operators.

§2. Stability under Group Actions

We now give some definitions and short explanations of the fundamental ideas. We first consider the situation of a finite dimensional Lie group G with corresponding Lie algebra L acting on a finite dimensional manifold M, with action $G \times M \to M$ denoted by

$$(g, m) \mapsto g.m \ .$$

The point $\mathring{m}$ is called *stable under the group action* G (or G-stable) if the orbit $G.\mathring{m} = \{g.\mathring{m} : g \in G\}$ contains a neighbourhood of $\mathring{m}$, i.e. for any point m in M sufficiently near $\mathring{m}$ there exists an element g in G such that

$$g.\mathring{m} = m. \tag{2.1}$$

For fixed $\mathring{m}$, the group action induces a map

$$F : g \to M$$

given by $F(g) = g.\mathring{m}$ with derivative

$$\phi : L \to M^\star$$

where we have introduced the notation $M^\star = T_{\mathring{m}} M$. Since G is a Lie group with Lie algebra L, ϕ is given by

$$\phi(x) = \frac{d}{dt}(g_t.\mathring{m})\big|_{t=0} \tag{2.2}$$

where g_t is the one-paramter subgroup corresponding to $x \in L$.

We say the point $\mathring{m}$ is *infinitesimally* G-*stable* if ϕ is onto, or equivalently if $\phi : L \to M^\star$ has a right inverse $\psi : M^\star \to L$.

It is not difficult to see that G-stability implies infinitesimal G-stability since given any $v \in M^\star$ and some curve $\gamma(t)$ in M with $\gamma'(0) = v$, stability allows us to define a corresponding curve $g(t)$ in G satisfying

$$\gamma(t) = g(t).\mathring{m}$$

and by taking the derivative of this equation at the origin we see that there exists an element $x = g'(0)$ in $T_e G = L$ such that $\phi(x) = \gamma'(0) = v$, whence $\phi : L \to M^\star$ is onto.

On the other hand by the surjective mapping theorem, infinitesimal G-stability implies G-stability so that the two conditions are equivalent in the finite dimensional case. Rather than appeal to the surjective mapping theorem we give an independent proof which can be adapted to the infinite dimensional case.

§3. Linearisation and Newton's tangent method.

Let $\alpha : U \to E$ be some chart for the manifold M in the neighbourhood U of $\mathring{m}$. Then we may use this chart to identify points in a neighbourhood of zero in $T_{\mathring{m}} M = M^*$ with points in a neighbourhood of $\mathring{m}$ by means of the map $f : U \to M^*$ given by

$$m^* := f(m) = \alpha_\star^{-1}(\alpha(m) - \alpha(\mathring{m})).$$

Now the linearised version of (2.1) is just

$$\phi(x) = m^*. \qquad (3.1)$$

In order to solve (2.1) an iterative procedure based upon repeatedly solving (3.1) is used. For consistency of notation in the iteration it is convenient to set $m_1 = m$ and $m_1^* = m^*$. Our first estimate is given by $g_1 = \exp(\psi(m_1^*))$, so that g_1 is an approximation to g, $g_1.\mathring{m}$ is close to m_1, and hence $g_1^{-1}.m_1$ is close to $\mathring{m}$. Let $m_2 = g_1^{-1}.m_1$, then we want to find an $h \in G$ such that

$$h.\mathring{m} = m_2 \qquad (3.2)$$

and the g required to solve (2.1) is given by $g = g_1 h$. Notice that (3.2) is just (2.1) with g replaced by h and m_1 replaced by m_2. Repeating this process, we obtain an approximation to h given by $g_2 = \exp(\psi(m_2^*))$ and $g_1 g_2$ is a better approximation to g than g_1. At the k-th stage of the iteration we have

$$g_k = \exp(\psi(m_k^*)), \qquad m_k = g_{k-1}.m_{k-1} \qquad (3.3)$$

and the estimate for g is

$$g_1 g_2 \cdots g_k. \qquad (3.4)$$

Note how the group structure has been used so that one needs the derivative of the group action at $\mathring{m}$ and not at any other point m.

Now since ψ is a map between finite dimensional vector spaces we have

$$\|\psi(m_k^\star)\| < C\|m_k^\star\|$$

where $C = \|\psi\|$. So by (3.3) we have

$$\|m_{k+1}^\star\| < C_1 \|m_k^\star\|^2. \tag{3.5}$$

The quadratic convergence of the sequence $\|m_k\|$ ensures that

$$g = \lim_{k\to\infty} g_1 g_2 \ldots g_k \tag{3.6}$$

exists and is a solution to (2.1). The above argument which shows that infinitesimal G-stability implies G-stability can be extended to the infinite dimensional case provided certain additional conditions are imposed.

§4. The infinite dimensional case: finite order and G-stability

For the applications we require, both $\mathcal{M}$ and $\mathcal{G}$ are infinite dimensional, and cannot be given the structure of Banach manifolds. The principal difficulty is that the existence of a right inverse ψ is no longer sufficient and additional conditions must be imposed to ensure that (3.6) conveges.

One such condition [4] is that ϕ have *finite order*. In the case where $\mathcal{M}$ is a space of analytic functions (or analytic vector fields) we say ϕ is of finite order ν if ϕ has a right inverse ψ (i.e. the point $\mathring{m}$ is infinitesimally $\mathcal{G}$-stable) such that for each function m analytic in a neighbourhood of some domain U, $\psi(m)$ is analytic in a domain V and for sufficiently small positive δ

$$\|\psi(m^\star)\|_V \le C\|m^\star\|_{U+\delta}\delta^{-\nu}$$

where C is a positive constant $\| \ \|_U$, $\| \ \|_V$ are suitable norms in $U \subset \mathcal{M}^\star$ and $V \subset \mathcal{L}$ and $U + \delta$ is U widened by an amount δ.

Now if one again uses the iterative approximation argument (Newton's Method) to solve the infinite dimensional version of (2.1), the "error" $m_k^\star$ at the k-th stage satisfies

$$\|m_{k+1}^\star\|_{U_k} \leq K\|\psi(m_k^\star)\|_{V_k}\|m_k^\star\|_{U_k} \qquad (4.1)$$

where $\|\ \|_{U_k}$ and $\|\ \|_{V_k}$ are suitable norms defined on domains $U_k \subset \mathcal{M}^\star$ and $V_k \subset \mathcal{L}$ respectively.

When ϕ is of finite order ν

$$\|m_{k+1}^\star\|_{U_k} \leq KC\|m_k^\star\|_{U_k - \delta}\delta^{-\nu}\|m_k^\star\|_{U_k}$$

and hence

$$\|m_{K+1}^\star\|_{U_k - \delta} \leq \tilde{C}\|m_k^\star\|_{U_k}^2 \delta^{-\nu}.$$

If we now set $U_{k+1} = U_k - \delta$ we have

$$\|m_{K+1}^\star\|_{U_{k+1}} \leq \tilde{C}\|m_k^\star\|_{U_k}^2 \delta^{-\nu},$$

so that by shrinking the domain by an amount δ it is possible to essentially maintain the quadratic convergence with the additional factor of δ^ν. By choosing $\delta = \delta_k$ appropriately at each stage of the iteration one can ensure that one still has the product $\lim_{k\to\infty} g_1 \ldots g_k$ converging to g but that the domain

$$U = \lim_{k\to\infty} U_k$$

does not shrink to nothing.

Thus in the infinite dimensional case one has the result that infinitesimal G-stability together with the finite order condition imply G-stability.(See for example the work of Zehnder [5] for a precise formulation of this result.) It is worth remarking that there are various other conditions used by other authors all of which have the effect of controlling the norm of the (approximate) inverse map and can be used to prove similar results. In particular the finite order condition is closely related to the condition used by R. S. Hamilton [6] that ϕ is a tame Fréchet operator between tame Fréchet spaces.

§5. Finite order, small divisors and exceptional sets

The condition that an operator be of finite order is closely related to the classical problem of small divisors, as can be seen from considering the example of the $\mathcal{G}$-stability of vector fields.

Let $\mathring{m}$ be the vector field on $\mathbf{C}^n$ given by

$$\mathring{m} = \sum_{k=1}^{n} \omega_k z_k \frac{\partial}{\partial z_k} \tag{5.1}$$

where $\omega_1, \ldots, \omega_n$ are constants. From now on, unless stated otherwise, k will always be an integer satisfying $1 \leq k \leq n$. Let $\tilde{\mathcal{M}}$ be the set of vector fields of the form $m = \mathring{m} + m^\star$ where $m^\star \in M^\star$ is of the form

$$m^\star = \sum_{k=1}^{n} f_k(z_1, \ldots, z_n) \frac{\partial}{\partial z_k} \tag{5.2}$$

$$\text{with} \qquad f_k(0) = 0 \qquad \text{and} \qquad \frac{\partial f_k}{\partial z_j}(0) = 0.$$

Let $\mathcal{G}$ be the (local) group of analytic homeomorphisms $g : \mathbf{C}^n \to \mathbf{C}^n$ with $g(0) = 0$ and Dg_0 the identity map with action $\mathcal{G} \times \tilde{\mathcal{M}} \to \tilde{\mathcal{M}}$ on $\tilde{\mathcal{M}}$ given by

$$g.m(z) = g_\star m(g^{-1}(z)). \tag{5.3}$$

The tangent space at the identity of the group of analytic homeomorphisms is the Lie algebra of analytic vector fields and the additional conditions that $g(0) = 0$ and that $Dg_0 = I$ ensure that the correponding vector fields satisfy $f_k(0) = 0$ and $\partial f_k / \partial z_j = O$. Hence $\mathcal{L} = T_e \mathcal{G} = \mathcal{M}^\star$.

For fixed $\mathring{m} \in \mathcal{M}$ we have the map

$$F : \mathcal{L} \to \mathcal{M}$$

with derivative

$$\phi : \mathcal{L} \to \mathcal{M}^\star$$

given by

$$\phi(x) = -\mathcal{L}_x \mathring{m} = [\mathring{m}, x] \qquad (5.4)$$

where $\mathcal{L}_x$ denotes the Lie derivative and $[\ ,\ ]$ denotes the Lie bracket.

Since $\mathcal{L} = \mathcal{M}^*$, ϕ is a linear operator from an infinite dimensional vector space into itself and we can find the eiegenfunctions and eigenvalues. We may use these to obtain an expression for the inverse operator when it exists and determine when ϕ satisfies the finite order condition.

It can be verified by direct calculation that for each $\mathbf{j} = (j_1,\ldots,j_n)$ with $j_k \geq 0$, $k = 1,\ldots,n$ and $j_1 + \cdots j_n \geq 2$, the vector fields $e_{\mathbf{j},k} : \mathbf{C}^n \to \mathbf{C}^n$ given by

$$e_{\mathbf{j},k}(z_1,\ldots,z_n) = z_1^{j_1}\ldots z_n^{j_n}\frac{\partial}{\partial z_k} \qquad (5.5)$$

form a complete set of eigenfunctions for ϕ and have corresponding eigenvalues $\mathbf{j}.\omega - \omega_k$.

For future convenience we define the set

$$\mathbf{J} = \{\mathbf{j} \in \mathbf{Z}^n : j_k \geq 0\ ,\ |\mathbf{j}|_1 \geq 2\}$$

where for each $\mathbf{v} \in \mathbf{R}^n, |\mathbf{v}|_1 = |v_1| + \cdots |v_n|$, so the set of eigenvalues of ϕ is

$$\{e_{\mathbf{j},k} : \mathbf{j} \in \mathbf{J},\ 1 \leq k \leq n\}.$$

We now use the eigenfunctions of ϕ to give an expression for the inverse function ψ. We first observe that any m^* in M can be written

$$m^* = \sum_{\substack{1 \leq k \leq n \\ \mathbf{j} \in \mathbf{J}}} c_{\mathbf{j},k}\, e_{\mathbf{j},k}.$$

Since $\phi(e_{\mathbf{j},k}) = (\mathbf{j}.\omega - \omega_k)e_{\mathbf{j},k}$, we have

$$\psi(e_{\mathbf{j},k}) = \frac{1}{(\mathbf{j}.\omega - \omega_k)}e_{\mathbf{j},k} \qquad (5.7)$$

and hence

$$\psi(m^*) = \sum_{\substack{1 \leq k \leq n \\ \mathbf{j} \in \mathbf{J}}} \frac{c_{\mathbf{j},k}}{(\mathbf{j}.\omega - \omega_k)}e_{\mathbf{j},k}. \qquad (5.8)$$

We can now see that a necessary condition for the existence of the right inverse ψ is that $(\mathbf{j}.\omega - \omega_k)$ does not vanish. We call the set

$$\{\mathbf{z} \in \mathbf{C}^n : \mathbf{j}.\mathbf{z} - z_k = 0, \; \mathbf{j} \in \mathbf{J}, \; 1 \le k \le n\}$$

the resonance set.

We therefore see that it is a necessary condition for a vector field to be infinitesimally $\mathcal{G}$-stable that the linearised vector field have non-resonant eigenvalues. On the other hand even if the eigenvalues are non-resonant, they may be close to resonance and the term $(\mathbf{j}.\omega - \omega_k)$ can become arbitrarily small for suitably large $|\mathbf{j}|$. This is the reason for calling such terms "small divisors". In order to control the norm of ψ so that the finite order condition is satisfied one must impose additional conditions which do not allow the eigenvalues to be "almost resonant" and ensure that $(\omega.\mathbf{j} - \omega_k)^{-1}$ does not grow too rapidly as $|\mathbf{j}|$ gets large. We can ensure that $\dot{m}$ is infinitesimally $\mathcal{G}$-stable and that ϕ is of finite order by demanding that $\omega \in \mathbf{C}^n$ satisfies

$$|\mathbf{j}.\omega - \omega_k| \ge C|\mathbf{j}|_1^{-\tau}, \qquad 1 \le k \le n, \qquad \mathbf{j} \in \mathbf{J} \qquad (5.9)$$

for some positive $\tau = \tau(\omega)$ and $C = C(\omega)$.

We now show how this condition implies finite order. For each positive number r let

$$\|m\|_r = \sup\{|f_k(z)| : z_1\bar{z}_1 + \cdots + z_n\bar{z}_n \le r^2, \; 1 \le k \le n\}$$

where

$$m^\star = \sum_{k=1}^n f_k(z)\frac{\partial}{\partial z_k}.$$

So by (5.6) we have for each k, $1 \le k \le n$, and $\mathbf{j} \in \mathbf{J}$

$$|c_{\mathbf{j},k}|r^{|\mathbf{j}|_1} \le \|m^\star\|_r. \qquad (5.10)$$

Hence for $0 < \rho < r < 1$ we have by (5.8)

$$\|\psi(m^*)\|_\rho \leq \sup\left\{ \left| \sum_{\mathbf{j} \in J} \frac{c_{\mathbf{j},k}}{(\mathbf{j}.\omega - \omega_k)} z_1^{j_1} \cdots z_n^{j_n} \right| : \right.$$

$$\left. z_1 \bar{z}_1 + \cdots + z_n \bar{z}_n \leq \rho^2,\ 1 \leq k \leq n \right\}$$

$$\leq \sup\{ \sum_{\mathbf{j} \in J} |c_{\mathbf{j},k}| |\mathbf{j}.\omega - \omega_k|^{-1} \rho^{|\mathbf{j}|_1} : 1 \leq k \leq n \}$$

$$\leq \|m^*\|_r \sum_{\mathbf{j} \in J} |\mathbf{j}.\omega - \omega_k|^{-1} \left(\frac{\rho}{r}\right)^{|\mathbf{j}|_1}$$

$$\leq C^{-1} \|m^*\|_r \sum_{\mathbf{j} \in J} |\mathbf{j}|_1^\tau \left(\frac{\rho}{r}\right)^{|\mathbf{j}|_1}$$

where we have used the Diophantine condition (5.9) to obtain the last inequality. The last sum can be estimated by $C_1(1-\rho/r)^{-\tau-n}$, so that

$$\|\psi(m^*)\|_\rho \leq C_2 \|m^*\|_r (r - \rho)^{-\tau-n} \qquad (5.11)$$

where C_2 is a positive constant. Putting $r = \rho + \delta$, $\nu = \tau + n$, we get

$$\|\psi(m^*)\|_\rho \leq C_2 \|m^*\|_{\rho+\delta} \delta^{-\nu} \qquad (5.12)$$

whence ϕ is of finite order, of at most $\nu = \tau + n$. The above proof that the Diophantine condition implies finite order involves only crude estimates and a more careful calculation using the methods of Rüssman [7] enables one to replace the exponent $\tau + n$ by τ. We also remark that the Diophantine condition ensures that (5.7) converges (in some domain U) and hence that $\hat{m}$ is infinitesimally $\mathcal{G}$-stable.

Let the set of ω satisfying (5.9) for a given τ and some $C > 0$ be $\Omega(\tau)$ i.e. let

$$\Omega(\tau) = \{ \mathbf{z} \in \mathbf{C}^n : \quad \text{for some } C > 0,$$

$$|\mathbf{j}.\mathbf{z} - z_k| \geq C|\mathbf{j}|_1, \mathbf{j} \in \mathbf{J}, 1 \leq k \leq n \}.$$

Then if $\omega \in \Omega(\tau)$, firstly no eigenvalues vanish and ϕ is onto so that $\hat{m}$ is infinitesimally $\mathcal{G}$-stable and secondly ϕ is of finite order, and hence $\dot{m}$ is $\mathcal{G}$-stable. For obvious reasons we call $\Omega(\tau)$ a $\mathcal{G}$-stability set. The complement $E(\tau) = \mathbf{C}^n \setminus \Omega(\tau)$ is given by

$$E(\tau) = \{\mathbf{z} \in \mathbf{C}^n : \quad \text{for each } C > 0,$$
$$|\mathbf{j}.\mathbf{z} - z_k| < C|\mathbf{j}|_1^{-\tau} \text{ for some } \mathbf{j} \in \mathbf{J}, 1 \leq k \leq n\}$$

and corresponds to vector fields $\dot{m}$ for which $\mathcal{G}$-stability cannot be proved. When $\tau > n/2 - 1$, $E(\tau)$ has Lesbegue measure zero and for this reason is called exceptional. It should be pointed out that the Diophantine condition we have imposed here is unnecessarilly restrictive and that one can establish the finite order requirement under somewhat weaker conditions. Indeed the most fruitful approach is to to work with the finite order condition directly and define the exceptional set to consist of those ω for which there exists a vector field which is not of finite order. Even with this approach there is room for improvement and one should probably replace the $\delta^{-\nu}$ power law requirement in the finite order condition by some weaker requirement which still allows one to prove that the iteration converges. The problem of finding the "true" exceptional set consisting of those ω for which there exists a $\mathcal{G}$-unstable vector field with ω as its eigenvalues is very difficult. However recent work by Yoccoz [8] has answered the analogous question for the problem of the linearisation of germs of analytic diffeomorphisms.

It is also worth pointing out that there are two distinct regions in which the vector ω may lie. The Siegel domain consisting of the set of points ω in $\mathbf{C}^n$ such that the origin lies in the convex hull of $\omega_1, \ldots, \omega_n$, and the Poincaré domain consisting of the set of points such that the convex hull does not contain the origin [9]. The resonant hyperplanes

$$H(\mathbf{j}, k) = \{\mathbf{z} \in \mathbf{C}^n : \mathbf{j}.\mathbf{z} = z_k\}$$

are everywhere dense in the Siegel domain, but occur discretely in the Poincaré domain. Thus the problem of "almost resonance" is only encountered in the Siegel domain.

§6. Cofinite $\mathcal{G}$-stability

So far we have discussed $\mathcal{G}$-stability of a point in terms of neighbourhoods. More generally, when considering infinite dimensional examples, an orbit through the point $\mathring{m}$ in $\mathcal{M}$ might not be open but might be of finite codimension, so that the orbit through $\mathring{m}$ and nearby orbits can be parameterised in some convenient way (see figure 1). A formal definition is that $\mathring{m}$ is *cofinite $\mathcal{G}$-stable* if there exists some finite dimensional subspace $\mathcal{M}_0$, a positive function $k(\epsilon)$ which tends to 0 with ϵ so that given any point m with $\rho(m, \mathring{m}) < \epsilon$ (where ρ is a metric on $\mathcal{M}$), there exists a $g(m)$ in $\mathcal{G}$ and a m' in $\mathcal{M}_0$ with $\rho(m, m') < k(\epsilon)$ and

$$g.m' = \mathring{m}. \tag{6.1}$$

The point $\mathring{m}$ in $\mathcal{M}$ is said to be *infinitesimally cofinite $\mathcal{G}$-stable* if the induced action of the Lie algebra satisfies the analogous condition, i.e. that the codimension of the image of the induced map $\phi : \mathcal{L} \to \mathcal{M}^\star$ is finite (so that ϕ has finite corank). As before when ϕ has finite order, these conditions are equivalent.

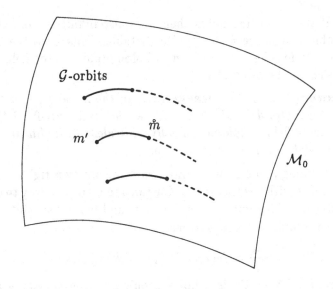

Figure 1

We now take $\mathcal{M}$ to consist of all vector fields

$$m = \sum_{k=1}^{n} f_k(z) \frac{\partial}{\partial z_k} \qquad (6.2)$$

where f is a holomorphic function (in a neighbourhood of the origin) satisfying $f(0) = 0$. The linear part $(m)_1$ of m is given by

$$(m)_1 = \sum_{j,k=1}^{n} a_{jk} z_j \frac{\partial}{\partial z_k}, \qquad (6.3)$$

where

$$a_{jk} = \frac{\partial f_k}{\partial z_j}(0, \dots, 0)$$

is invariant under the action of $\mathcal{G}$. The vector field $\dot{m}$ will be cofinite stable if every nearby vector field with the same linear part lies in the orbit $\mathcal{G}.\dot{m}$. In this case we take $\mathcal{M}_0$ to be the space of $n \times n$ matrices (a_{jk}).

The condition that ϕ has finite order is still the condition that the eigenvalues $\omega = (\omega_1, \dots, \omega_n)$ lie in $\Omega(\tau)$ for some positive τ, i.e. for all $\mathbf{j} \in \mathbf{J}$ and all $k = 1, \dots, n$, ω satisfies the inequality

$$|\omega.\mathbf{j} - \omega_k| \geq C|\mathbf{j}|^{-\tau} \qquad (6.4)$$

for some $C > 0$.

If this condition holds then $\dot{m}$ is also infinitesimally cofinite $\mathcal{G}$-stable and hence $\dot{m}$ is cofinite $\mathcal{G}$-stable. This gives the following result for the normal form of holomorphic vector field which vanishes at the origin.

Theorem 6.1. *If the eigenvalues ω of the linear part of a holomorphic vector field which vanishes at the origin satisfy (6.4), then the vector field is holomorphically equivalent to its linear part in some neighbourhood of the origin.*

Although this result is interesting in its own right, one can combine it with the theorem of Cartan and Carathéodory to obtain a criterion which guarantees the (past and future) stability of the system of differential equations

$$\frac{dz_k}{dt} = f_k(z_1, \dots, z_n) \qquad k = 1, \dots, n \qquad (6.5)$$

where $f : \mathbf{C}^n \to \mathbf{C}^n$ is a holomorphic function vanishing at the origin.

Theorem 6.2. *If $\frac{\partial f_k}{\partial z_j}(0,\ldots,0)$ is diagonalisable with purely imaginary eigenvalues $\omega_k = i\alpha_k$, $\alpha_k \in \mathbf{R}$ and if for some $C > 0$ and $\tau > 0$,*

$$|\alpha.\mathbf{j} - \alpha_k| \geq C|\mathbf{j}|^{-\tau}$$

for all $\mathbf{j} \in \mathbf{J}$ and $1 \leq k \leq n$, then the solution $z_k(t) = 0$, $k = 1,\ldots,n$ is a stable solution to (6.5).

§7. Normal forms and Siegel's Theorem

In this section we investigate the normal form for a holomorphic homeomorphism in the neighbourhood of a fixed point and consider the stability of the fixed point under iteration by the map to both the future and the past. For a holomorphic homeomorphism $f : \mathbf{C}^n \to \mathbf{C}^n$, with $f(0) = 0$, we say that $z = 0$ is a (past and future) stable fixed point if for every neighbourhood U of 0 there exists a neighbourhood V with $0 \in V \subset U$ such that $\hat{z} \in V$ implies $f^k(\hat{z}) \in U$ for all $k \in \mathbf{Z}$.

In order to study the normal form of holomorphic homeomorphisms in the neighbourhood of a fixed point, we now take $\mathcal{M}$ to be the set of homeomorphisms

$$f : \mathbf{C}^n \to \mathbf{C}^n$$

such that $f(0) = 0$, which are holomorphic in some neighbourhood of the origin. We take $\mathcal{G}$ to be the group of homeomorphisms

$$g : \mathbf{C}^n \to \mathbf{C}^n$$

such that $g(0) = 0$, $Dg_0 = I$, which are holomorphic in some neighbourhood of the origin, and consider the action $\mathcal{G} \times \mathcal{M} \to \mathcal{M}$ of $\mathcal{G}$ on $\mathcal{M}$ given by

$$(g, f) \mapsto g.f = g \circ f \circ g^{-1}.$$

We say that two maps f_1, $f_2 \in \mathcal{M}$ are biholomorphically equivalent if there exists a $g \in \mathcal{G}$ such that

$$f_2 = g.f_1.$$

Given a homeomorphism $f \in \mathcal{M}$, the linear part $(f)_1$ is the linear map given by

$$(f)_1 : z \mapsto Df_0(z).$$

As in the case of vector fields we will be interested in finding those maps f which are (biholomorphically) equivalent to their linear part.

We could again look at the induced map between tangent spaces and investigate the condition that this be of finite order. But rather than study the normal form for a holomorphic diffeomorphism in a neighbourhood of a fixed point (which we may take to be the origin) directly we establish a relationship between the normal form for such a function and the normal form for a system of holomorphic differential equations where the coefficients can also be periodic functions of time t. We start with the system of differential equations

$$\frac{dz}{dt} = Az + Q(z,t) \tag{7.1}$$

where $Q : \mathbf{C}^n \times \mathbf{R} \to \mathbf{C}^n$ satisfies $Q(z, t+2k\pi) = Q(z,t)$ for $k \in \mathbf{Z}$ and

$$Q(0,t) = 0 \qquad \frac{\partial Q_k}{\partial z_j}(0,t) = 0.$$

The solutions of (7.1) can be used to define a holomorphic map $f : \mathbf{C}^n \to \mathbf{C}^n$ (at least in a neighbourhood of the origin) by setting $f(\mathring{z}) = z(2\pi)$ where $z(t)$ is the unique solution to (7.1) with the initial condition $z(0) = \mathring{z}$. The conditions on Q ensure that the origin is a fixed point. If we think of a solution to (7.1) as a flow on $\mathbf{C}^n \times S^1$, then f is just its Poincaré map.

Given any smooth diffeomorphism $f : \mathbf{R}^n \to \mathbf{R}^n$ with fixed-point the origin, by considering its suspension one can construct a vector field on $\mathbf{R}^n \times S^1$ whose flow has Poincaré map f. Although such a vector field may not be smooth it may be "smoothed" without changing the Poincaré map. We thus obtain a differential equation with periodic coefficients for which the diffeomorphism f is the Poincaré map. The problem is more complicated in the holomorphic case, where the problem is equivalent to showing the holomorphic triviality of holomorphic foliations over S^1. Nevertheless one can show that given any holomorphic diffeomorphism

$f : \mathbf{C}^n \to \mathbf{C}^n$ with fixed point the origin it is still possible to construct a system of differential equations of the form (7.1) whose Poincaré map is f [9].

In order to deduce results for the normal form and stability of f we first give the periodic versions of Theorem 6.1 and Theorem 6.2 which give results for the normal form and stability of a vector field on $\mathbf{C}^n \times S^1$ with constant linear part.

Theorem 7.1. *If the eigenvalues $\omega = (\omega_1, \ldots, \omega_n)$ of the linear part of the vector field satisfy for some $C > 0$ and $\tau > 0$*

$$|\omega.\mathbf{j} + iq - \omega_k| \leq C|\mathbf{j}|_1^{-\tau}$$

for all $\mathbf{j} \in \mathbf{J}$, $q \in \mathbf{Z}$ and $k = 1, \ldots, n$, then it is possible to introduce new coordinates ζ by a holomorphic transformation, 2π-periodic in t, such that (7.1) becomes

$$\frac{d\zeta}{dt} = A\zeta. \tag{7.2}$$

Theorem 7.2. *If A is diagonalisable with purely imaginary eigenvalues $\omega_k = i\alpha_k$, $\alpha_k \in \mathbf{R}$ and if for some $C > 0$ and $\tau > 0$*

$$|\alpha.\mathbf{j} + q - \omega| \leq C|\mathbf{j}|^{-\tau}$$

for all $\mathbf{j} \in \mathbf{J}$, $q \in \mathbf{Z}$ and $k = 1, \ldots, n$, then the solution $z_k(t) = 0$, $k = 1, \ldots, n$ is a stable solution to (7.1).

In the ζ-coordinates given by Theorem 7.1 the associated Poincaré map $f : \mathbf{C}^n \to \mathbf{C}^n$ for the section $t = 0$ is given by

$$f(\zeta) = B\zeta \tag{7.3}$$

where B is the matrix given by $B = \exp 2\pi A$. So that the transformation $z \mapsto \zeta$ (at $t = 0$) has the effect of linearising f.

We can rewrite the condition on the eigenvalues of A in terms of the eigenvalues $\beta_1, \ldots, \beta_n$ of B as

$$\left| \beta_k - \prod_{m=1}^{n} \beta_m^{j_m} \right| \geq C'|\mathbf{j}|^{-\tau} \tag{7.4}$$

for some $C' > 0$ and $\tau > 0$ and for all $\mathbf{j} \in \mathbf{J}$ and $k = 1, \ldots, n$.

On the other hand the transformation from z to ζ coordinates has derivative equal to the identity at the origin so that $B = Df_0$ is nothing more than the Jacobian of the original map f at the origin. We therefore have the following theorem for the normal form for a holomorphic map in the neighbourhood of a fixed point.

Theorem 7.3. *Let $f : \mathbb{C}^n \to \mathbb{C}^n$ be a holomorphic map with fixed point the origin. Let $\beta_1, \ldots, \beta_n$ the eigenvalues of $\frac{\partial f_k}{\partial z_j}(0, \ldots, 0)$ satisfy condition (7.4). Then in some neighbourhood of the origin f is biholomorphically equivalent to its linear part.*

The above theorem also provides conditions for the past and future stability of the fixed point under iteration by f.

Theorem 7.4. *Let $f : \mathbb{C}^n \to \mathbb{C}^n$ be a holomorphic map with fixed point the origin. Let $\beta_1, \ldots, \beta_n$ the eigenvalues of the Jacobian at the origin satisfy condition (7.4) and in addition have modulus one. Then the origin is a past and future stable fixed point under iteration by f.*

§8. References

[1] A. Liapounoff, 'Problème général de la stabilité du mouvement' *Ann. Fac. Sci. Toulouse* **2** (1907) 203-474 reprinted in *Ann. Math. Studies* **17**.

[2] S. Bochner and W. T. Martin, *Several Complex Variables*, (Princeton, 1948).

[3] M. R. Herman, *Recent results and some open questions on Siegel's linearization theorem of germs of complex analytic diffeomorphisms of $\mathbb{C}^n$ near a fixed point*, Preprint (1988).

[4] V. I. Arnol'd, *Singularity Theory* (LMS Lecture Note Series 53 C.U.P Cambridge 1981).

[5] E. Zehnder, ' Generalised implicit function theorems with applications to some small divisors problems I', *Comm. Pure and App. Math.*, **28** (1975) 91–140.

[6] R. S. Hamilton, 'The inverse function theorem of Nash and Moser', *Bull. A.M.S.*, **7** (1982) 65–223.

[7] H. Rüssmann, 'Über invariant Kurven differenzierbarer Abbildungen eines Kreisringes', *Nachr. Akad. Wiss. Göttingen II, Math. Phys.Kl.* (1970) 67–105.

[8] J. C. Yoccoz, 'Siegel's theorem for quadratic polynomials' Preprint (1988).

[9] V. I. Arnol'd, *Geometrical methods in the theory of ordinary differential equations*, (Springer-Verlag, New-York 1983).

3
Metric Diophantine approximation of quadratic forms

S.J.Patterson

Georg-August-Universität, Göttingen, FRG

§1. Introduction

Since Artin's paper [1] the modular group in association with the techniques of Diophantine approximation has served as a model for the study of dynamical systems related to discrete groups acting on hyperbolic space. Several contributions to this conference deal with such questions. There are however many more arithmetic groups acting on hypererbolic space than the modular group and in this essay I shall explain the number-theoretic significance of some of the finer results. The translation is based on well-known principles and the knowledgable reader will find little that is new here. This discussion is expository; its purpose is to explain this translation on the basis of certain examples and to make the methods more generally available, for there is no particularly satisfactory account from the point of view we shall be taking here.

First of all we shall introduce $(N + 1)$-hyperbolic space in the Poincaré model $B^{N+1} = \{\mathbf{x} \in \mathbf{R}^{N+1} : ||\mathbf{x}||^2 < 1\}$. We define the group $\mathrm{Con}(N)$ of diffeomorphisms of B^{N+1} which preserve the metric

$$ds^2 = ||d\mathbf{x}||^2/(1 - ||\mathbf{x}||^2)^2$$

For $\mathbf{x}_1, \mathbf{x}_2 \in B^{N+1}$ we let

$$L(\mathbf{x}_1, \mathbf{x}_2) = \frac{1}{2} + \frac{||\mathbf{x}_1 - \mathbf{x}_2||^2}{(1 - ||\mathbf{x}_1||^2)(1 - ||\mathbf{x}_2||^2)}$$

which is related to the hyperbolic distance $d(\mathbf{x}_1, \mathbf{x}_2)$ between $\mathbf{x}_1$ and $\mathbf{x}_2$ by

$$\cosh 2d(\mathbf{x}_1, \mathbf{x}_2) = 2L(\mathbf{x}_1, \mathbf{x}_2).$$

In particular if $g \in \mathrm{Con}(N)$ then

$$L(g(\mathbf{x}_1), g(\mathbf{x}_2)) = L(\mathbf{x}_1, \mathbf{x}_2).$$

We have to discuss the geometrical meaning of this construction. Let $c \in \mathbf{R}$ and consider the hyperboloid

$$Q(c) = \{(\mathbf{x}, y) \in \mathbf{R}^{N+1} \times \mathbf{R} : y^2 - ||\mathbf{x}||^2 = c\}.$$

If $c > 0$ this is two-sheeted, if $c < 0$ it is one sheeted and if $c = 0$ it is a cone. The group $O(N + 1, 1)$ acts on each $Q(c)$; if $c \neq 0$ then the action is transitive and it is also transitive on $Q(0) \setminus \{0\}$. If $c > 0$ then a subgroup of index 2 preserves each of the two sheets. We can identify this subgroup with $\mathrm{Con}(N)$ as follows. For $c \geq 0$ we let $Q^+(c) = \{(\mathbf{x}, y) \in Q(c) : y > 0\}$. The map $p : Q^+(c) \to B^{N+1}$ defined by

$$p(\mathbf{x}, y) = \mathbf{x}/(y + \sqrt{c})$$

(the stereographic projection) is a bijection and if $c > 0$

$$L(p(\mathbf{z}_1), p(\mathbf{z}_2)) = 1/2c(y_1 y_2 - \mathbf{x}_1 \cdot \mathbf{x}_2)$$

where $\mathbf{z}_i = (\mathbf{x}_i, y_i)$, $i = 1, 2$ and "$\cdot$" denotes the Euclidean inner product. Thus the induced action of the subgroup $O^+(N + 1, 1)$ of $O(N + 1, 1)$ of index 2 which preserves all the $Q^+(c)$ acts on B^{N+1} to preserve L. It is a theorem of Liouville's that $\mathrm{Con}(N)$ is no larger. We shall identify $\mathrm{Con}(N)$ with $O^+(N+1, 1)$ henceforth.

If $c > 0$ the stabiliser of a point of $Q(c)$ is isomorphic to $O(N+1)$ and this is a maximal compact subgroup of $O^+(N+1, 1)$. If $c < 0$ then the stabiliser of a point is isomorphic to $O(N, 1)$; this is realised by the induced action on the tangent space which inherits a quadratic form. The stabiliser of a point of $Q(0) \setminus \{0\}$ is isomorphic to a semi-direct product of $O(N)$ by $\mathbf{R}^N$. These are refered to as the *elliptic, hyperbolic* and *parabolic* cases respectively.

We shall investigate certain arithmetic subgroups of $\mathrm{Con}(N)$. Denote the quadratic form $y^2 - ||\mathbf{x}||^2$ on $\mathbf{R}^{N+1} \times \mathbf{R}$ by $q(\mathbf{z})$ for $\mathbf{z} = (\mathbf{x}, y)$. Let Λ be a lattice in $V = \mathbf{R}^{N+1} \times \mathbf{R}$ on which q takes integral values. Let Γ be the subgroup of $O(N+1, 1)$ which preserves Λ. It is known that Γ acts discontinuously on $Q(1)$ and that the quotient has finite volume. If q takes on the value 0 on $\Lambda \setminus \{0\}$ then the quotient is not compact ; in this case $(Q(0) \setminus \{0\}) \cap \Lambda \neq \emptyset$ and if $\mathbf{z}$ belongs to this set then $\{g \in \Gamma : g(\mathbf{z}) = \mathbf{z}\}$ is cocompact in $\{g \in O(N+1,1) : g(\mathbf{z}) = \mathbf{z}\}$. Moreover the group Γ has only finitely many orbits on $p((Q(0) \setminus \{\mathbf{0}\}) \cap \Lambda) \subset S^N$ — but not on $Q(0) \setminus \{\mathbf{0}\}$. Likewise for any $c \neq 0$, Γ has only finitely many orbits on $Q(c) \cap \Lambda$. We shall study the distribution of these orbits as subsets of $Q(c)$. Recall that by Meyer's theorem ([2], Cor. 1 to Theorem 6.1.1) if $N \geq 3$ then q will represent 0 on $\Lambda \setminus \{\mathbf{0}\}$. Before we do this some remarks are in order. First of all it is somewhat more usual to assume that $\Lambda = \mathbf{Z}^{N+2}$ and that q is an arbitrary quadratic form of signature $(1, N + 1)$ with integer coefficients. The reader should find no difficulty in reformulating the results from this viewpoint. The second point is that there are more general arithmetic subgroups of $O(N+1,1)$ than those defined above. The more general construction is as follows. Let k be a totally real number field and let Σ_∞ be the set of real places of k, that is of embeddings of k into $\mathbf{R}$. There are $[k : \mathbf{Q}]$ elements of Σ_∞. Let v_0 be one element of Σ_∞, to be kept fixed. Let k_∞ be $\mathbf{R}^{[k:\mathbf{Q}]}$ and we consider k embedded in k_∞ by $\prod v$, i.e. for each element of Σ_∞ we consider the coresponding embedding into $\mathbf{R}$. Let q be a quadratic form on k^{N+2} considered also as one on k_∞^{N+2}. Let R be the ring of integers of k. Let Λ be a lattice in k^{N+2}, that is a finitely generated R-submodule which generates k^{N+2} as a vector space. Then Λ is cocompact in k_∞^{N+2}. Let $O(q, k_\infty)$ be the orthogonal group of q over the ring k_∞, that is the product of the orthogonal groups of the real quadratic forms obtained from q by the embeddings $v \in \Sigma_\infty$ of k into $\mathbf{R}$. Let Γ be the subgroup of $O(q, k_\infty)$ which preserves Λ. This is again a discrete subgroup of $O(q, k_\infty)$ with finite covolume. If q is of signature $(1, N + 1)$ in k_{v_0} and is definite in all other real completions of k then the projection of $O(q, k_\infty)$ onto $O(q, k_{v_0})$ has a compact kernel and the image of Γ in $O(q, k_{v_0})$ is discrete and of finite covolume; it is also isomorphic to Γ. Since $O(q, k_{v_0})$ is isomorphic to $O(N+1,1)$ the group Γ can be considered as a discrete subgroup of this group with finite covolume. We shall not discuss this case in detail here but remark that it is not difficult to extend our considerations to this case.

For completeness we note that the connected component of $O(2,1)$ is isomorphic to $PSL_2(\mathbf{R})$. The isomorphism can be realised by observing that if X denotes the three-dimensional space of 2×2 symmetric matrices then $PSL_2(\mathbf{R})$ acts on X by $g : A \rightarrow gA^tg$. The determinant is a quadratic form on X which is preserved under this action ; it is of signature $(1,2)$. The modular group then is the subgroup preserving *the subset of integral matrices in X*. Also the connected component of $O(3,1)$ can be identified with $PSL_2(\mathbf{C})$, where one realises the isomorphism by considering the action of this group on the set of complex 2×2 matrices A such that $^tA = \overline{A}$ given by $g : A \rightarrow gA^t\overline{g}$, again equipped with the determinant as the quadratic form.

We shall now recall some results on diophantine approximation in discrete subgroups of $\mathrm{Con}(N)$ of finite covolume. These are taken from [5]. The results in [5] are proved only in the case $N = 1$; however either the proofs can be extended without much difficulty or we can refer to [3] and [8] for closely related results, and there are only minor gaps remaining. Let henceforth Γ be a dicrete subgroup of $\mathrm{Con}(N)$ with finite covolume. We recall that by a theorem of Thurston's ([9], Prop. 8.4.3) Γ is geometrically finite, that is it has a fundamental polyhedron with finitely many faces. This is what is needed in [5]. The first pair of theorems we shall recall are the following:

Theorem 1. *Suppose that Γ has parabolic elements and let P_0 be a set of representatives of the cusps of Γ. Then there exists a constant $c > 0$ so that if $\zeta \in S^N$ and if $X > 1$ then there exist $g \in \Gamma$ with $L(0, g(0)) < c.X$ and $p \in P_0$ with*

$$||g(p) - \zeta||^2 < (X.L(0, g(0)))^{-1}.$$

Theorem 2. *Suppose that Γ has no parabolic elements and let $H = \{\eta, \eta'\}$ be the set of fixed points of a hyperbolic subgroup of Γ. Then there exists a constant $c > 0$ so that if $\zeta \in S^N$ then there exist $g \in \Gamma$ with $L(0, g(0)) < c.X$ and $\eta_1 \in H$ with*

$$||g(\eta_1) - \zeta||^2 < X^{-2}.$$

These theorems are generalizations of Dirichlet's theorem on diophantine approximation. See [5] §7. They form the basis of the later theorems. The proofs of these two theorems are simple exercises using the same circle of ideas as Hedlund's Lemma. From them we can deduce the following metric theorem:

Theorem 3. *Let w be a positive decreasing function on $[2, \infty)$ for which there exists a $c > 0$ so that $w(2x) > c.w(x)$. Let $\mathbf{y}$ be a parabolic vertex if there are any, and a hyperbolic fixed point otherwise. Let $A(\mathbf{y})$ be the set of $\mathbf{x} \in S^N$ for which there exist infinitely many $g \in \Gamma$ with*

$$\|\mathbf{x} - g(\mathbf{y})\| < w(L(0, g(0)))/L(0, g(0)).$$

Then $A(\mathbf{y})$ is of zero Lebesgue N-measure if for some $K > 1$

$$\sum_{n \geq 1} w(K^n)^N$$

converges; otherwise $A(\mathbf{y})$ is of full measure in S^N.

For this see [5] §9. Note that the convergence condition is independent of K. This theorem is based on the well-known theorem of Khinchin. The following result is an extension of a theorem of Jarník ; the proof given in [5] §10 does not extend easily to the general case but an independent proof in the case that A (as in the formulation of the theorem) consists of parabolic vertices has been given by Dani in [3]. For the purposes of our applications the only case for which there does not exist a proof in the literature is when $N = 2$ and A consists of hyperbolic fixed points. We shall leave the proof in this case to the conscientious reader.

Theorem 4. *Let A be a finite set of parabolic cusps of Γ if there are any, and a finite set of hyperbolic fixed points if there are none. Let E be the set of points $\mathbf{x} \in S^N$ for which there exists a constant $c(\mathbf{x}) > 0$ so that for all $g \in \Gamma$ and all $\mathbf{y} \in A$ one has*

$$\|g(\mathbf{y}) - \mathbf{x}\| > c(\mathbf{x})/L(0, g(0)).$$

Then E has Hausdorff dimension N.

Observe that in these theorems

$$L(\mathbf{x}, g(\mathbf{x}'))/L(0, g(0))$$

considered as a function of g on $\mathrm{Con}(N)$ is bounded above and below by functions of $\mathbf{x}$, $\mathbf{x}'$ alone. Thus the seemingly arbitrary nature of $L(0, g(0))$ is explained by the fact that in the theorems any of the other $L(\mathbf{x}, g(\mathbf{x}'))$ could have been used as a measurement of the size of g. In [5] $L(0, g(0))$ is denoted by $\mu(g)$.

We shall now interpret these in terms of the quadratic form q and the lattice Λ introduced above. Theorem 1 yields:

Theorem 5. *Let q, Λ be as above. Suppose that $Q^+(0) \cap \Lambda \neq \emptyset$. Then there exists $C > 0$ so that for each $\zeta \in Q^+(0)$ and each $X > 1$ there exists $\mathbf{x} \in Q^+(0) \cap \Lambda$, $\|\mathbf{x}\| < C.X$ and*

$$0 \leq q(\zeta, \mathbf{x}) \leq \|\zeta\|/X,$$

where $q(.,.)$ is the symmetric bilinear form so that $q(\mathbf{x}, \mathbf{x}) = q(\mathbf{x})$.

Theorem 6. *Let q, Λ be as above. Suppose that $Q(0) \cap \Lambda \neq \emptyset$. Let $w : [1, \infty) \to \mathbf{R}_+^\times$ be a decreasing positive function so that $\inf w(2\mathbf{x})/w(\mathbf{x}) > 0$. Then the set*

$$\{\zeta \in Q(0) : \text{there exist infinitely many } \mathbf{x} \in Q(0) \cap \Lambda$$
$$\text{so that } 0 \leq q(\zeta, \mathbf{x}) \leq w(\|\mathbf{x}\|)^2 \|\zeta\|/\|\mathbf{x}\|\}$$

has zero or full N-dimensional Lebesgue measure in S^N according to whether $\sum_{n \geq 1} w(K^n)^N$ converges or diverges for some $K > 1$.

Theorem 7. *Let q, Λ be as above. Suppose that $Q^+(0) \cap \Lambda \neq \emptyset$. Then the set*

$$\{\zeta \in Q(0) : \text{there exists } \epsilon > 0 \text{ so that for all}$$
$$\mathbf{x} \in Q(0) \cap \Lambda \quad |q(\zeta, \mathbf{x})| > \epsilon.\|\zeta\|/\|\mathbf{x}\|\}$$

has Hausdorff dimension $N + 1$ but is of zero $(N + 1)$-measure.

The translations from Theorems 1,3 and 4 to Theorems 5, 6 and 7 is fairly straightforward and we sketch it merely in the case of Theorem 5. The proofs of Theorems 6 and 7 will be left to the reader.

The basis of the translation is the notion of a horocycle. Let $\zeta \in S^N$ and $\mathbf{w} \in B^{N+1}$; define the Poisson kernel

$$P(\mathbf{w}, \zeta) = (1 - \|\mathbf{w}\|^2)/\|\mathbf{w} - \zeta\|^2.$$

Then we define

$$H(\zeta, r) = \{\mathbf{w} \in B^{N+1} : P(\mathbf{w}, \zeta) = r\}$$

which is a horocycle at ζ. Let $c > 0$; then one checks that for $\mathbf{y} \in Q^+(0)$ one has

$$H(p(\mathbf{y}), r) = p(\{\mathbf{x} \in Q^+(c) : q(\mathbf{x}, \mathbf{y}) = (\sqrt{c}/2r).\|\mathbf{y}\|\}).$$

Now suppose that ζ is a cusp of Γ and denote by $\Gamma(\zeta)$ the stabiliser of ζ in Γ. Then there exist constants c_1, $c_2 > 0$ so that if ζ is fixed and r is large enough then for any $g \in \Gamma$ there exists $g' \in g.\Gamma(\zeta)$ so that if $g(H(\zeta, r)) = H(g(\zeta), r')$ then

$$L(\mathbf{0}, g'(\mathbf{0})) < c_1.r'/r$$

whereas

$$L(\mathbf{0}, g(\mathbf{0})) > c_2.r'/r.$$

This is a simple geometrical lemma — see [5] Lemma 4.7. An immediate consequence of this and the description of a horocycle in terms of q given above is that there exist c_3, $c_4 > 0$ so that if $\mathbf{y} \in Q^+(0)$ then for $g \in \Gamma$ one has $\|g(\mathbf{y})\| < c_3 L(\mathbf{0}, g(\mathbf{0}))\|\mathbf{y}\|$ and there exists $g' \in g.\Gamma(\mathbf{y})$ so that $\|g'(\mathbf{y})\| > c_4 L(\mathbf{0}, g'(\mathbf{0}))\|\mathbf{y}\|$. Here we are regarding Γ as a subgroup of $O(N + 1, 1)$ and $\Gamma(\mathbf{y})$ is the stabiliser of $\mathbf{y}$; note that this group is in our case the same as $\Gamma(p(\mathbf{y}))$ in the notation above, since we can choose $\mathbf{y}$ to be primitive in Λ. We can now prove Theorem 5. In Theorem 1 we can replace the $g(p)$ by $p(\mathbf{x})$ for some $\mathbf{x} \in Q^+(0) \cap \Lambda$ and we can take $L(\mathbf{0}, g(\mathbf{0}))$ to be comparable with $\|\mathbf{x}\|$, and this gives the optimal choice of g. If we write $g(p) = p(\mathbf{x})$, $\zeta = p(\mathbf{z})$ then a simple calculation shows that

$$\|p(\mathbf{x}) - p(\mathbf{z})\|^2 = 4.q(\mathbf{x}, \mathbf{z})/\|\mathbf{x}\|.\|\mathbf{z}\|$$

from which the assertion follows immediately.

The considerations in the case where $Q(0) \cap \Lambda = \{0\}$ are more subtle. In this case we shall approximate elements of $Q(0)$ by 'hyperbolic pairs'. In other words we shall seek orthogonal decompositions $U \oplus H$ of $V = \mathbf{R}^{N+2}$ and $M \oplus M^\star$ of Λ where the orthogonal direct sum is with respect to q, $q|U$ is positive definite and $q|H$ is of signature $(1, 1)$. Also $M = \Lambda \cap U$, $M^\star = \Lambda \cap H$. We recall that the discriminant of a lattice Λ with respect to a symmetric bilinear form q is defined to be $\det(q(\mathbf{e}_i, \mathbf{e}_j))$ where $\mathbf{e}_i$ form a free basis of Λ. We write the discriminant of Λ with respect to q as $\mathrm{discr}_q(\Lambda)$. We have

$$\mathrm{discr}_q(\Lambda) = \mathrm{discr}_{q|U}(M).\mathrm{discr}_{q|H}(M^\star).$$

Moreover if $q|\Lambda$ is integral then $\mathrm{discr}_q(\Lambda)$ is an integer. Let $D = -\mathrm{discr}_q(\Lambda)$. Note that

$$1 \leq -\mathrm{discr}_{q|H}(M^*) \leq D.$$

Let Γ_0 be the subgroup $O(N+1,1)$ stabilising Λ. If $g \in \Gamma_0$ and $U \oplus H$ is a decomposition as above then $g(U) \oplus g(H)$ is another decomposition of $\mathbf{R}^{N+1}$ with $\Lambda = g(M) \oplus g(M^*)$. It is a classical consequence of the reduction theory of quadratic forms (see, for example [2], Chap.9, Proof of Lemma 6.1) that there are only finitely many inequivalent classes of decomposition with respect to this construction considered as an equivalence relation.

We shall seek next to investigate the structure of M^*. Let $d = -\mathrm{discr}_{q|H}(M^*)$. Let $\mathbf{f}_1, \mathbf{f}_2$ be a basis of M^* ; then

$$q(\mathbf{f}_1,\mathbf{f}_1)q(\mathbf{f}_2,\mathbf{f}_2) - q(\mathbf{f}_1,\mathbf{f}_2)^2 = -d.$$

We seek $\eta = \alpha\mathbf{f}_1 + \beta\mathbf{f}_2$ so that $q(\eta) = 0$. Then α, β would have to satisfy the homogeneous quadratic equation

$$\alpha^2 q(\mathbf{f}_1,\mathbf{f}_1) + 2\alpha\beta q(\mathbf{f}_1,\mathbf{f}_2) + \beta^2 q(\mathbf{f}_2,\mathbf{f}_2) = 0.$$

If d were a square then there would exist integral solutions of this equation which would contradict our initial assumption that q does not represent 0 on Λ. Thus d is not a square and we can find η, η' defined over $\mathbf{Q}(\sqrt{d})$ so that $q(\eta) = 0$, $q(\eta') = 0$ and η and η' are conjugate to one another.

Let now $\mathbf{e}_1,\ldots,\mathbf{e}_N$ be a basis of U so that

$$\|\mathbf{e}_i\| = 1, \qquad \mathbf{e}_i.\mathbf{e}_j = 0, \quad i \neq j,$$

$$q(\mathbf{e}_i) = \lambda_i, \qquad q(\mathbf{e}_i,\mathbf{e}_j) = 0, \quad i \neq j,$$

where $\mathbf{x}.\mathbf{y}$ denotes the Euclidean inner product. This is possible by the principal axes theorem. With respect to q the $\mathbf{e}_i$ are orthogonal to the η, η' but not necessarily with respect to the Euclidean inner product. We shall now show that all but one of the λ_i are equal to 1. Let $\mathbf{x}^-$ be the projection of $\mathbf{x}$ onto the factor $\mathbf{R}$ of $V = \mathbf{R}^{N+1} \times \mathbf{R}$. Then one has

$$q(\mathbf{x}) + 2(\mathbf{x}^-)^2 = \|\mathbf{x}\|^2$$

We shall assume that $\mathbf{e}_i$ is normalised by $\mathbf{e}_i^- \geq 0$. Then it follows that $\mathbf{e}_i^- = ((1 - \lambda_i)/2)^{1/2}$. Then the fact that $\mathbf{e}_i$ are orthogonal with respect to q and the Euclidean inner product means that

$$(1 - \lambda_i)(1 - \lambda_j) = 0, \qquad i \neq j$$

from which it follows that at most one of the λ_i is not equal to 1. We order the $\mathbf{e}_i$ so that this is λ_N which we now denote by λ. Next $\mathbf{e}_i.\eta = q(\mathbf{e}_i, \eta) + 2.(\mathbf{e}_i^-)\eta^- = 2.(\mathbf{e}_i^-)\eta^-$. This is zero if $i \neq N$ and it is $\sqrt{2}.(1 - \lambda)^{1/2}$ if $i = N$. An analogous argument shows that

$$\eta^- = ||\eta||/\sqrt{2}.$$

Likewise one sees that

$$q(\eta, \eta') + ||\eta||.||\eta'|| = \eta \cdot \eta'.$$

Let us assume now that the η, η' are normalised by $||\eta||, ||\eta'|| = 1$ (so they are no longer necessarily defined over $\mathbf{Q}(\sqrt{d})$). Then we have enough information to calculate the determinants of q and E (the Euclidean inner product) with respect to the basis $\mathbf{e}_1, \ldots \mathbf{e}_N, \eta, \eta'$ in terms of λ and $q(\eta, \eta')$. These have to differ by a factor of -1 and this yields a relation between λ and $q(\eta, \eta')$. In fact we find

$$q(\eta, \eta') = -2\lambda/(\lambda + 1).$$

This leads to the relation

$$||\eta - \eta'||^2 = -4q(\eta, \eta').$$

Suppose now that $\zeta \in Q^+(0)$, $||\zeta|| = 1$. Then one verifies after a short calculation that

$$||\zeta - \eta||.||\zeta - \eta'||/||\eta - \eta'|| = [q(\zeta, \eta)q(\zeta, \eta')/(-q(\eta, \eta'))]^{1/2}.$$

Let us now turn to the approximation theorems. We shall consider choices of splittings $U \oplus H$ of V and $M^* \oplus M$ of Λ. These give rise to a pair of points η, η' on S^N as we have shown above. The stabiliser of this pair in Γ is cocompact in the stabiliser in $\mathrm{Con}(N)$. We denote these two stabilisers by Γ_H and $\mathrm{Con}(N)_H$ respectively. We observe also that U is the orthogonal complement of H with respect to q and so we can denote it by U_H. The possible Γ_H belong to finitely many conjugacy classes in Γ. If we consider H to be fixed then for $g \in \Gamma$ the group $g\Gamma_H g^{-1}$ fixes $g(\eta)$, $g(\eta')$ and there are only a finite number of conjugates $g\Gamma_H g^{-1}$ of Γ_H such that $\|g(\eta) - g(\eta')\| > r$ for any $r > 0$. Moreover by [5] Lemma 4.8 there exists a constant $c > 0$ so that if such a conjugate is given then there exists $g_1 \in \Gamma$ so that $g_1\Gamma_H g_1^{-1} = g\Gamma_H g^{-1}$ and $L(0, g(0)) < c.\|g(\eta) - g(\eta')\|^{-1}$. Now the pair $g(\eta)$, $g(\eta')$ corresponds to the splitting $gU \oplus gH$. In view of the discussion above we see that

$$\mathrm{discr}_{q|U}(M)/\mathrm{discr}_{E|U}(M) = \lambda$$

where E is the Euclidean inner product and

$$\|\eta - \eta'\|^2 = 8\lambda/(\lambda + 1).$$

Since $1 \leq \mathrm{discr}_{q|U}(M) \leq D$ we can use $\mathrm{discr}_{E|M}(M)$ as a measure of $\|\eta - \eta'\|^{-2}$. With this remark we can now formulate the number-theoretic forms of theorems 2,3 and 4.

Theorem 8. *Suppose that q does not represent 0 on Λ. Then there exist a constant $C > 0$ so that if $\zeta \in Q(0) \setminus \{0\}$ is given then for any $X > 1$ there exists a decomposition of Λ as $M^\star \oplus M$ and of $\mathbf{R}^{N+1} \times \mathbf{R}$ as $H \oplus U$ where H is a hyperbolic plane with $M^\star = \Lambda \cap H$ and U is positive definite with $M = \Lambda \cap U$ so that if $\mathrm{pr}_H(\zeta)$ is the orthogonal projection of ζ into H with respect to q then*

$$0 < q(\mathrm{pr}_H(\zeta)) < X^{-1}$$

and if E represents the Euclidean quadratic form then

$$\mathrm{discr}_{E|U}(M) < C.X.$$

Theorem 9. *Suppose that q does not represent 0 on Λ. Let w be a real function satisfying the conditions of Theorem 3. Then the set of $\zeta \in Q(0)$ for which there exists a splitting $V = U \oplus H$, $\Lambda = M \oplus M^\star$ as above for which*

$$0 < q(\mathrm{pr}_H(\zeta)) < w(\mathrm{discr}_{E|U}(M))^2/\mathrm{discr}_{E|U}(M)$$

is of zero or full $(N+1)$-*measure according to whether* $\sum\limits_{n\geq 1} w(K^n)^N$
converges or diverges for a $K > 1$.

Theorem 10. *Suppose that* q *does not represent* 0 *on* Λ. *Suppose that* $N = 1$. *Then the set of* $\zeta \in Q(0)$ *for which there exists* $\epsilon > 0$ *so that for all splittings* $V = U \oplus H$, $\Lambda = M \oplus M^*$ *one has*

$$q(\mathrm{pr}_H(\zeta)) > \epsilon/\mathrm{discr}_{E|U}(M)$$

has Hausdorff dimension $N+1$ *but is of zero* $(N+1)$-*dimensional measure.*

Finally we shall deal with a topic which is closely related to this and is more closely related to ergodic-theoretic methods than one might expect. This concerns the distribution of points of $Q(c) \cap \Lambda$ on $Q(c)$. The methods developed in [4] and [6] can be applied directly to this case and we have the folowing theorem:

Theorem 11. *With the notations above we have that if* m *is a non-zero integer and* $\mathbf{w} \in Q(1)$ *then*

$$\mathrm{Card}\{\lambda \in \Lambda : q(\lambda) = m, \; |q(\lambda, w)| < X\} = c.g(m).X^N + O(X^\alpha)$$

where $\alpha < N$ *and* $g(m)$ *is the measure of the representation of* m *by the form* q *on* Λ *in the sense of [7] Chap.IV, §6 and C is a constant depending on* N.

It is worth noting that in the language of this paper Siegel's theta function for the quadratic form q on Λ is essentially the following expression defined for z in the upper half-plane and $\eta \in Q(1)$

$$\Theta_\Lambda(z, \eta) = \sum_{\lambda \in \Lambda} e^{2\pi i (z.q(\lambda) + i.\mathrm{Im}(z).q(\lambda, \eta)^2)}.$$

Siegel in [7] Chap.IV has shown that this satisfies a transformation law under a congruence subgroup of the modular group as a function of z, and also that the 'average of this' in η over a fundamental domain on $Q(1)$ for the action of Γ is an Eisenstein series, at least if $N > 2$. The Dirichlet series

$$\sum_{\substack{\lambda \in \Lambda \\ q(\lambda) = m}} |q(\lambda, \eta)|^{-2s}$$

can be represented as the inner product of the corresponding m-th Poincaré-Petersson series from which Theorem 10 can be deduced. This method is related via Hecke theory to the method of [6] since

$$e^{-yq(\eta,\eta')^2}$$

as a function of η, η' is a point-pair invariant for Con(N) on $Q(1)$.

References

[1] E. Artin, 'Ein mechanisches System mit quasiergodischen Bahnen', *Abh. Math. Sem. Hamburg* (1924) 170–175.

[2] J. W. S. Cassels, *Rational Quadratic Forms*, (Academic Press, 1978).

[3] S. G. Dani, 'Bounded orbits of flows on homogeneous spaces', *Comment. Math. Helvet.*, **61** (1986) 636–700.

[4] P. Lax and R. Phillips, 'The asymptotic distribution of lattice points in euclidean and non-euclidean spaces', *J. Fnl. Analysis* **46** (1982) 280–350.

[5] S. J. Patterson, 'Diophantine approximation in Fuchsian groups', *Phil. Trans. Royal Soc. London A* **262** (1976) 527–563.

[6] S. J. Patterson, 'A lattice point problem in hyperbolic space', *Mathematika* **22** (1975) 81–88.

[7] C. L. Siegel, *Lectures on quadratic forms*, (Tata institute 1956).

[8] D. Sullivan, 'Discrete conformal groups and measurable dynamics', *Proc. Symp. Pure Math.* **39** (1984) 169–185.

[9] W. P. Thurston, *The geometry and topology of 3-manifolds*, (Notes, Princeton, 1983).

4
Symbolic dynamics and Diophantine equations
Caroline Series
University of Warwick, Coventry, UK

§1. The problems

Certain classical Diophantine problems have a geometrical interpretation as the height to which geodesics travel up the cusp of the modular surface $H/SL(2, Z)$, where $H = \{z \in C : \operatorname{Im} z > 0\}$ is the hyperbolic plane and $SL(2, Z)$ acts by linear fractional transformations. My purpose here is to show how the apparently rather imprecise methods of symbolic dynamics not only suggest generalisations of the classical results, but also carry very detailed and precise numerical information. More precisely, I consider the following geometrical problem:

Problem A. Let γ be a geodesic in H and let $\operatorname{ht}(\gamma) = 1/2|\gamma^+ - \gamma^-|$, where γ^+, γ^- denote the positive and negative endpoints of γ on R. (The choice of orientation of γ at this point is irrelevant but the notation $\gamma^\pm$ will be convenient later.) Let Γ be a zonal Fuchsian group acting on H, that is a Fuchsian group containing a translation $W : z \to z + \lambda, \lambda \in R$. Find

$$\text{e.h.}(\gamma) = \sup\{\operatorname{ht}(g\gamma) : g \in \Gamma\}$$

in particular, find those γ for which e.h.(γ) is small.

We call e.h.(γ) the *essential height* of γ. When γ is projected onto H/G it measures the height which γ travels up the cusp corresponding to W.

In the case $\Gamma = SL(2, Z)$, this problem is equivalent to the following well known classical questions:

Problem B. Minima of binary indefinite quadratic forms
Let $Q(x, y) = ax^2 + bxy + cy^2$ be an indefinite binary quadratic form with $a, b, c \in Z$ and let $\Delta_Q = b^2 - 4ac$. Find

$$\inf\left\{|Q(x, y)|\Delta_Q^{-1/2} : (x, y) \in Z^2 \backslash \{(0, 0)\}\right\}.$$

Problem C. Diophantine Approximation For $\xi \in \mathbf{R} \setminus \mathbf{Q}$, find

$$c(\xi) = \inf\{c \in \mathbf{R} : |\xi - p/q| < c/q^2 \text{ for infinitely many } q \in \mathbf{N}\}$$

The equivalence of **B** and **C** when ξ is a quadratic surd is well known, see for example [2]. The direct relation of **C** to **A** has been explained in various places such as [9,6].

To see the connection of **B** with **A** one associates to an indefinite quadratic form Q the geodesic γ_Q joining the roots of $Q(\xi, 1)$ on **R**. Then

$$\mathrm{ht}(\gamma_Q) = 1/2\Delta_Q^{1/2} Q(1,0)^{-1},$$

so that by comparing the several actions of $G = SL(2, \mathbf{Z})$ on **H**, on quadratic forms, and on $\mathbf{Z}^2$ one obtains

$$(2\sup\{\mathrm{ht}(g\gamma_Q) : g \in G\})^{-1} = \inf\{(gQ)(1,0)\Delta_{gQ}^{-1/2} : g \in G\}$$

$$= \inf\{Q(g1, g0)\Delta_Q^{-1/2} : g \in G\}$$

$$= \inf\{|Q(x, y)|\Delta_Q^{-1/2} : x, y \in \mathbf{Z}\setminus\{0\}\}.$$

The classical theory of problems **B** and **C** is due to Markoff [8] and can be found for example in Cassels [2]. The corresponding solution to the geometrical version **A** in the case $\Gamma = SL(2, \mathbf{Z})$ originated with H. Cohn [4] and has appeared recently with various generalisations and refinements in [1,5]. In its simplest form it appears as follows:

Theorem 1. *Let γ be a geodesic in* **H** *and let $G = SL(2, \mathbf{Z})$. Then e.h.$(\gamma) \leq 1/3$ if and only if γ projects to a simple geodesic on* **H**$/[G, G]$ *with equality if and only if γ is not closed.*

Here a simple geodesic is one with no proper self intersections. Such geodesics may be open or closed. We note that $H = [G, G]$ is the free group on two generators

$$A = \begin{pmatrix} 1 & 1 \\ 1 & 2 \end{pmatrix}, \qquad B = \begin{pmatrix} 1 & -1 \\ -1 & 2 \end{pmatrix}$$

and that $M = $ **H**$/H$ is a punctured torus which forms a six sheeted covering of the modular surface **H**$/G$, see Figure 2a. The simple geodesics on M are precisely the images under $\mathrm{Aut}(H)$ of the axes

Ax(A), Ax(B) of A and B,which are of course themselves simple geodesics. Restricting to orientation preserving automorphisms (recall that every automorphism of H may be realised as an isometry of M) we have Aut(H) $= SL(2, \mathbf{Z})$ and the action of Aut(H) on homology $H_1(M)$ is the natural action of $SL(2, \mathbf{Z})$ on $\mathbf{Z}^2$. (This action of $SL(2, \mathbf{Z})$ should not be confused with the action by linear fractional transformations on H). The action of $U \in$ Aut(H) on G is determined (up to conjugation) by the fact that each homology class contains a *unique* simple geodesic; thus, for each $(p, q) \in \mathbf{Z}^2$ there is, up to cyclic permutation, exactly one cyclically reduced word $w(A, B) \in H$ whose abelianisation is $A^p B^q$.

The simple words (that is the words corresponding to simple closed geodesics), are conveniently arranged in a tree Θ corresponding to the decomposition of elements of $SL(2, \mathbf{Z}) =$ Aut(H) as products of upper and lower triangular matrices. The tree is shown in Figure 1. Of course there are really four such trees $\Theta(A, B)$, $\Theta(A, B^{-1})$, $\Theta(A^{-1}, B)$ and $\Theta(A^{-1}, B^{-1})$ corresponding to the four possible choices of initial generating pair. The tree illustrated is $\Theta(A, B)$. The automorphisms Δ_α, Δ_β which map a point $(\alpha, \beta) \in \Theta$ to the two points $(\alpha, \alpha^{-1}\beta)$ and $(\beta^{-1}\alpha, \beta)$ directly below are given by the matrices $\begin{pmatrix} 1 & 0 \\ -1 & 1 \end{pmatrix}$ and $\begin{pmatrix} 1 & -1 \\ 0 & 1 \end{pmatrix}$. Geometrically, they represent the Dehn twists about the axes of α and β. Every pair $(\alpha, \beta) \in \Theta$ is a possible pair of generators for H, and every possible pair appears [3]. In the classical Diophantine problems B and C, Θ appears in the guise of the *Markoff tree* in which the generating pairs (α, β) are replaced by integer triples (x, y, z) satisfying $x^2 + y^2 + z^2 = 3xyz$. In fact $3x = \operatorname{tr}\alpha$, $3y = \operatorname{tr}\beta$ and $3z = \operatorname{tr}\alpha\beta$. The above equation is exactly the trace identity

$$\operatorname{tr}^2 \alpha + \operatorname{tr}^2 \beta + \operatorname{tr}^2 \alpha\beta = \operatorname{tr}\alpha \operatorname{tr}\beta \operatorname{tr}\alpha\beta$$

(here we use the fact that $\operatorname{tr}[\beta^{-1}, \alpha^{-1}] = -2$). The Markoff moves

$$(x, y, z) \rightarrow (x, 3xy - z, y), \qquad (x, y, z) \rightarrow (3xy - z, x, y)$$

obtained from the trace identity $\operatorname{tr} uv = \operatorname{tr} u \operatorname{tr} v - \operatorname{tr} uv^{-1}$, are the Dehn twists Δ_α, Δ_β described above.

The words appearing in the tree Θ are characterised in yet another way in [11].

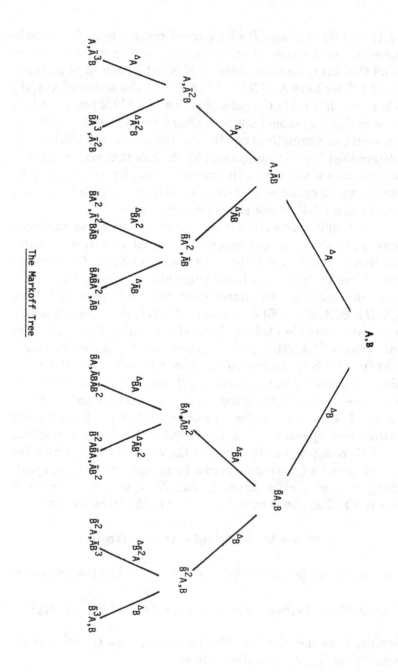

The Markoff Tree

Figure 1

We shall be concerned with investigating the analogue of Theorem 1 when $SL(2, \mathbf{Z})$ is replaced by the Hecke group G_5. This is the group generated by $z \to z + \omega$ and $z \to -1/z$ where $\omega = (1 + \sqrt{5})/2$ is the golden mean.

By a result of Leutbecher [7] this is equivalent to replacing $\mathbf{Q}$ by $\mathbf{Q}(\sqrt{5})$ in problem **C**.

Note: Actually our statement of this result is not quite accurate, because of the non-uniqueness of the representation of $\xi \in \mathbf{Q}(\sqrt{5})$ as p/q where p and q are integers in $\mathbf{Q}(\sqrt{5})$ with no common factor. In fact p and q may be multiplied by a unit, which in $\mathbf{Q}(\sqrt{5})$ may be arbitrarily large. Thus as it stands the formulation of Problem **C** makes no sense. Rather, one should write

$$c(x) = \inf\{c : |x - g(\infty)| < c/(c(g))^2\}$$

for infinitely many distinct cusps $g(\infty)$, where

$$g = \begin{pmatrix} a(g) & b(g) \\ c(g) & d(g) \end{pmatrix}.$$

It appears to be an interesting problem to detect which of the possible representations of $\xi \in \mathbf{Q}(\sqrt{5})$ corresponds to $a(g)/c(g)$, $g \in G_5$. This latter representation is unique. This can be done either by using the continued fractions of Rosen [9] or by implementing Leutbecher's proof [7], but one would prefer a more direct method.

Our aim is to find an analogue of the Markoff tree for G_5 and to show that the geodesics represented by this tree have small essential heights. There is an immediate objection to this programme, for by a result of Beardon et al. [1] a dichotomy of the type described in Theorem 1 between simple and non-simple curves can only occur on a punctured torus or a 3- or 4-times punctured sphere. On any other punctured surface, there are non-simple curves far away from the cusps. Since the analogue of the punctured torus M above is a punctured genus two surface, Beardon's result will apply and the dichotomy cannot persist. This can also be seen easily from the symbolic dynamics, as we shall explain below.

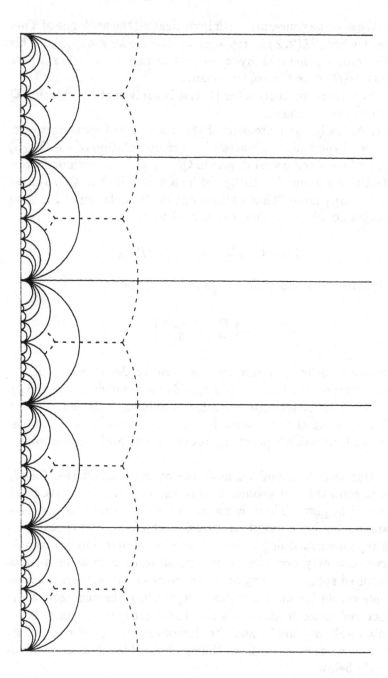

Figure 2a

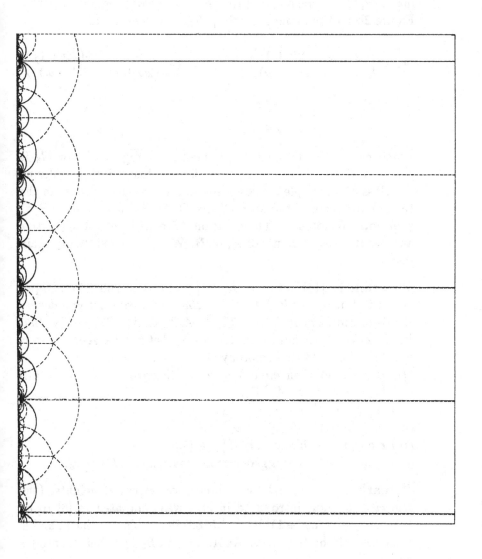

Figure 2b

The punctured surface M_5 in question is a ten sheeted covering of $\mathbf{H}/G_5$, corresponding to the fundamental regions shown in Figure 2b and fundamental group H_5 with generators

$$A_5 = \begin{pmatrix} 2+\omega & 2\omega+1 \\ 2\omega+1 & 2(1+\omega) \end{pmatrix} \qquad B_5 = \begin{pmatrix} 2+\omega & -(2\omega+1) \\ -(2\omega+1) & 2(1+\omega) \end{pmatrix}$$

$$C_5 = \begin{pmatrix} 1 & \omega \\ \omega & 2+\omega \end{pmatrix} \qquad D_5 = \begin{pmatrix} 1 & -\omega \\ -\omega & 2+\omega \end{pmatrix}$$

which pair opposite sides. A fundamental region Q_5 for G_5 is shown in dotted lines on the same figure. Instead of considering the class of all simple geodesics on M_5, we restrict to those in the the two branches of the Markoff tree $\Theta(A_5, B_5)$ and $\Theta(A_5^{-1}, B_5^{-1})$ henceforth denoted $\underline{\Theta}$. The notation $A_5^{\pm}$ etc. denotes the positive and negative fixed points of A_5 on $\mathbf{R}$. We can then state our main result.

Theorem 2. [14] *Let* $\nu_0 = B_5^{-1}(A_5^+) - A_5^- = 1.2059885\ldots, \nu_1 = B_5^{-1}((B_5^{-1}A_5)^+) - (B_5^{-1}A_5)^- = 1.2059896\ldots$ *and let* $\underline{\Theta}_C$ *denote words of the form* $A_5 B_5^{-n}$, $A_5^n B_5^{-1}$, $A_5^{-n} B_5$, $A_5^{-1} B_5^n$, $n \in \mathbf{N}$, *that is, the outer branches of the two trees. Let* γ *be a geodesic in* $\mathbf{H}$ *and let* $[\gamma]$ *denote its homotopy class in* H_5. *Then:*

(i) if $[\gamma] \in \underline{\Theta}$ *then* $e.h.(\gamma) \leq \nu_1$, *with equality if and only if* $[\gamma] = B_5^{-1} A_5$ *or* $A_5^{-1} B_5$.

(ii) if $[\gamma] \notin \underline{\Theta}$ *then* $e.h.(\gamma) > \nu_0$.

(iii) $\inf\{e.h.(\gamma) : [\gamma] \notin \underline{\Theta}\} = \nu_0$.

(iv) $e.h.(\gamma) < \nu_0$ *if and only if* $[\gamma] \in \underline{\Theta}_C$.

(v) $\inf\{e.h.(\Lambda) : \Lambda$ *is a geodesic lamination on* $M_5\} = \nu_0$.

Remark 1. A global lower bound on essential heights, (the *Hurwitz constant* introduced in [9]), was computed in [6] to be $1/2((1 - \cos^2 \pi/5)^2 + 1)^{-1/2} = 1.0180789\ldots$. This is sharp, and is achieved only by the curves Ax A_5 and Ax B_5 and their translates under G_5.

Remark 2. The laminations refered to in (v) are limits of sequences of geodesics in $\underline{\Theta}$ obtained by moving down branches of the tree. Every path down the tree defines a particular lamination. All the leaves in such a lamination have the same essential height, see [11,14].

Remark 3. The subgroup H_5^* of H_5 generated by A_5 and B_5 is free and the commutator $[B_5^{-1}, A_5^{-1}]$ is hyperbolic. Thus H/H_5^* is a one-holed torus. There is a formula due to Haas [5] for computing the distance of approach of a simple curve on such a surface to the geodesic bounding the hole, which in this case is the projection on H/H_5^* of $\mathrm{Ax}([B_5^{-1}, A_5^{-1}])$. This gives an estimate for the upper bound of essential heights of geodesics in $\underline{\Theta}$. One obtains $1.20614\ldots$ as opposed to $1.20598\ldots$ in Theorem 2(ii).

  Notation: From now on we shall write $G = G_3 = SL(2, \mathbf{Z})$ and $H = H_3 = [G_3, G_3]$,

$$A = A_3 = \begin{pmatrix} 1 & 1 \\ 1 & 2 \end{pmatrix}$$

and

$$B = B_3 = \begin{pmatrix} 1 & -1 \\ -1 & 2 \end{pmatrix}.$$

When there is no need to distinguish between G_3 and G_5, A_3 and A_5 etc, we shall drop subscripts and write G, A, etc...

§2. The Proofs

The proof of the first half of Theorem 1 is not difficult.

Lemma 2.1. *Suppose that Γ is a Fuchsian group containing a translation $W : z \to z + \lambda$, $\lambda \in \mathbf{R}$. Let γ be a geodesic on $\mathbf{H}$ which projects to a simple geodesic $\pi(\gamma)$ on $\mathbf{H}/\Gamma$. Then $e.h.(\gamma) \leq \lambda/2$.*

Proof. Suppose that to the contrary, we may find $\gamma \in G$ so that $ht(g\gamma) > \lambda/2$. Then $g\gamma \cap W(g\gamma) \neq \emptyset$ and projecting onto $\mathbf{H}/\Gamma$ we see that $\pi(\gamma)$ cannot be simple.

  In the classical case of G_3 the subgroup H_3 contains the commutator $[B_3^{-1}, A_3^{-1}] : z \to z+6$. Thus if γ is simple $e.h.(\gamma) \leq 3$. To improve the bound to $3/2$ as in Theorem 1 we proceed as follows. Let $J_3 \in \mathrm{Aut}\, H_3$ be defined by $J_3(A_3) = A_3^{-1}$, $J_3(B_3) = B_3^{-1}$. Viewed as an isometry on the fundamental region R_3 shown in Figure 2(a), J_3 is rotation by π about $\mathrm{Ax}\, A_3 \cap \mathrm{Ax}\, B_3 = i$. As isometries of $\mathbf{H}$ we have $J_3 A_3 = A_3^{-1} J_3$ and $J_3 B_3 = B_3^{-1} J_3$. Since J_3 is in the centre of $\mathrm{Aut}(H_3) = \mathrm{Isom}(M_3)$, and since every simple geodesic is the image of $\mathrm{Ax}\, A_3$ or $\mathrm{Ax}\, B_3$ under an isometry of M_3, we obtain that $J_3(\gamma) = \gamma^{-1}$ for all simple geodesics

γ. (Here γ^{-1} denotes the geodesic γ with orientation reversed.)
Now let T_3 be the isometry $B_3^{-1}J_3A_3$, so that $T_3^2 = [B_3^{-1}, A_3^{-1}]$
and $T_3 : z \to z + 3$. Since T_3 only differs from J_3 by covering
translations we see on projecting to M_3 that $\pi(T_3\gamma) = \pi(\gamma^{-1})$, in
other words, $\pi(T_3\gamma)$ and $\pi(\gamma)$ are identical as point sets on M_3.
But then the argument of Lemma 2.1 applies to give the bound
$e.h.(\gamma) \le 3/2$.

In the case of G_5 the idea is to replace T_3 by $T_5 = B_5^{-1}J_5A_5$,
where J_5 is rotation by π about $\text{Ax}\,A_5 \cap \text{Ax}\,B_5 = i$ (so that in fact
$J_5 = J_3$). We find that $T_5^2 = [B_5^{-1}, A_5^{-1}]$ is now hyperbolic and so
has a well-defined sqare root:

$$T_5 = \begin{pmatrix} -(3\omega + 2) & -(4\omega + 3) \\ -3\omega & -(3\omega + 2) \end{pmatrix}$$
$$= \begin{pmatrix} -6.85410197\ldots & -9.47213596\ldots \\ -4.85410187\ldots & -6.85410197\ldots \end{pmatrix}$$

One computes easily that

$$\sup_{x \in \mathbf{R}}(T(x) - x) = 2.4120183\ldots$$

Arguments identical to those given above for G_3 therefore give an
upper bound of $1.206009\ldots$ for curves in Θ.

It will be seen that this is not as sharp as the statement of
Theorem 2(i).

For the proof of the remaining part of Theorem 1 and Theorem 2 we introduce symbolic dynamics. Since we are dealing with
essential heights relative to the group G_3 or G_5 (henceforth written G) and not H_3 or H_5 (henceforth written H), it is convenient
to introduce symbols which are invariant relative to these larger
groups, but which nevertheless make use of the nice fact that that
H tesselates $\mathbf{H}$ by ideal polygons. (An ideal polygon is a polygon
with all its vertices at infinity). This always makes life much easier as explained at length in [13]. In the case of H_3 the symbols
we shall describe are intimately related to continued fractions [11,
12]. The symbols for H_5 have a translation into the continued
fractions for $\mathbf{Q}(\sqrt{5})$ developed by Rosen in [10].

In both cases we divide the fundamental region R for H in
half by the imaginary axis and denote the right half by S. Letting

G act on S we obtain a tesselation $\mathcal{T}(S)$ of **H** by ideal triangles or pentagons as the case may be. The subgroup of G fixing one of these triangles or pentagons is cyclic of order 3 or 5, and conjugate to that generated by the elliptic element $\left(\begin{smallmatrix} 1 & -1 \\ 1 & 0 \end{smallmatrix}\right)$ or $\left(\begin{smallmatrix} \omega & -1 \\ 1 & 0 \end{smallmatrix}\right)$ respectively. This element rotates S about $e^{\pi i/3}$, respectively $e^{\pi i/5}$. Any geodesic which crosses one of the regions S^* in $\mathcal{T}(S)$ (for simpicity we omit geodesics which end in a cusp, although they can be included in the discussion without difficulty) joins a pair of sides of S^*. We label this segment n, $n \in \{1,2\}$, respectively $n \in \{1,2,3,4\}$, if the side across which it enters S^* is separated from the side across which it leaves by $n - 1$ sides of S^*, counting clockwise from the entering side. This is illustrated in Figure 3. Clearly, these symbols are invariant under the action of G. Hence to any geodesic γ (not terminating in a cusp), one associates a doubly infinite sequence $\sigma(\gamma) = \{\sigma(\gamma)_n\} \in \prod_{-\infty}^{\infty}\{1,2\}$, respectively $\prod_{-\infty}^{\infty}\{1,2,3,4\}$, where $\sigma(\gamma)_n$ denotes the type of the n-th segment counted from some arbitrarily chosen initial point. It is not hard to show that precisely because S has all its vertices on **R**, every sequence in the product space determines a unique geodesic in **H** once the initial side is chosen. Different choices of initial side correspond to the images $g\gamma$, $g \in G$, of γ under G. All such choices project to the same geodesic on **H**/G. We refer to $\sigma(\gamma)$ as the *σ-sequence* of γ.

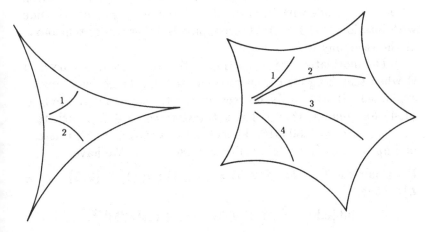

Figure 3 — Symbolic Coding

We shall also need another symbolic dynamics which relates directly to words in the groups H. Symbols of this kind have been described at length in [13]. Here we only need the simplest case. Each side of the fundamental region h for H is labelled by that generator of H which identifies it with the opposite side of R, as shown in Figure 2. This labelling is transported by the action of of H to the whole tesselation $T(R)$. A geodesic in **H** is labelled by the sequence $\chi(\gamma) = \{\chi(\gamma)_n\} \in \prod_{-\infty}^{\infty} F$, where F is the generating set $\{A_3^{\pm 1}, B_3^{\pm 1}\}$ or $\{A_5^{\pm 1}, B_5^{\pm 1}, C_5^{\pm 1}, D_5^{\pm 1}\}$ as the case may be. The sequence $\chi(\gamma)$ is called the χ-*sequence* of γ. It is not hard to show, again using the fact that R is an ideal polygon, that a doubly infinite *reduced* sequence in the generators F uniquely determines a geodesic in **H** once an initial choice has been made. In this way a closed geodesic on $M = \mathbf{H}/H$ is represented by an infinite periodic sequence, the fundamental period of which is a cyclically reduced word representing its homotopy class in H, and an open geodesic by a doubly infinite aperiodic reduced sequence in F. We shall speak of cyclically reduced words and infinite reduced periodic words interchangeably.

There is an obvious coding between χ- and σ-sequences, because the χ-labelling of a geodesic arc crossing a region $R^{\star}$ of $T(R)$ uniquely determines its σ-sequence. Thus for example in the case of G_5 we have $CD \to 11$, $C\overline{D} \to 2$, $A\overline{D} \to 21$. (Notice that our convention is that the χ-sequence XY denotes a segment entering $R^{\star}$ across the side with interior label X and leaving across the side with interior label Y^{-1}.) This mapping is either six- or ten-to-one as the case may be.

The motivation for restricting to the class of simple curves in $\underline{\Theta}$ when searching for curves of small essential height can now be explained. It can easily be seen in the G_5 case that if $\sigma(\gamma)_n \in \{1, 4\}$ for some n, then γ has a translate $g\gamma$, $g \in G_5$, such that $ht(g\gamma) > \omega/2 = 0.809016$. In fact, as was shown in [6], this can be improved to $(1 + \cos^2 \pi/5)^{1/2} = 1.286276\ldots$. We have:

Proposition 2.2. *Let* $\Sigma(2,3) = \{\gamma \subset \mathbf{H} : \sigma(\gamma)_n \in \{2,3\} \quad \forall n \in \mathbf{Z}\}$. *Then*

$$\inf\{e.h.(\gamma) : \gamma \notin \Sigma(2,3)\} \geq (1 + \cos^2 \pi/5)^{1/2}.$$

It is not hard to see that many non-simple curves are contained in $\Sigma(2,3)$, while many simple ones are not. Thus Proposi-

tion 2.2 can be regarded as a rather concrete version of Beardon's
result [1] in this particular case. Further, one sees that although
$AB^{-1} = 22$, $A^2 = 23$ and $B^{-2} = 32$ one has $\sigma(AB) = 4$ so that
$\Sigma(2,3)$ cannot form a subgroup of H_5. This is the reason that the
trees $\Theta(A_5, B_5^{-1})$, $\Theta(A_5^{-1}, B_5)$ are excluded from Theorem 2.
Pre-Simple Curves. We need to translate the condition that a
word $w(A,B) \in H$ belongs to Θ into a condition on σ-sequences.
It is convenient to begin with a large class of words which we call
(for want of a better name) *presimple*. These are words of the
form $AB^{-n_1} AB^{-n_2} \ldots AB^{-n_k}$, $n_i > 0$, and the analogous forms
obtained from the other branch of Θ and by taking inverses. The
following lemma is easily checked.

Lemma 2.3. *If $w(A,B)$ is presimple then its σ-sequence lies in
$\Sigma(2,3)$ if $G = G_5$ and contains none of the following blocks:*
  (i) X^3
  (ii) $X^2YX \ldots XYX^2$
  (iii) $XYX^2Y^2 \ldots Y^2X^2YX$,

*where in the case of G_3, $\{X,Y\} = \{1,2\}$ and in the case of G_5,
$\{X,Y\} = \{2,3\}$.*

Notice that the converse of Lemma 2.3 is false; for example a
σ-sequence may be positioned so that its χ-sequence contains the
symbol C.

To recognize simple words one considers the effect of the two
Dehn twists $\Delta_\alpha : (\alpha,\beta) \to (\alpha, \alpha^{-1}\beta)$ and $\Delta_\beta : (\alpha,\beta) \to (\beta^{-1}\alpha,\beta)$,
$\alpha, \beta \in \Theta$ which represents progress down the branches of the tree.
Suppose that $w = AB^{-n_1} AB^{-n_2} \ldots AB^{-n_k}$ is presimple and that
$n = \min\{n_i : 1 \le i \le k\}$.
  Define

$$A^{(1)} = B^{-n} A, \qquad B^{(1)} = B.$$

Then

$$(A^{(1)}, B^{(1)}) = \Delta_B^n(A, B),$$

so that $A^{(1)}$, $B^{(1)}$ is a generating pair; further when $A^{(1)}$, $B^{(1)}$ are
substituted for A and B in w to form a word

$$w^{(1)} = w^{(1)}(A^{(1)}, B^{(1)})$$

there are no cancellations so that $w^{(1)}$ has length strictly less than
w. If $w^{(1)}$ is again presimple the operation can be repeated. We
call $w^{(1)}$, $w^{(2)}, \ldots$ the *derived* words of w.

Proposition 2.4. [3,14] *A word $w(A, B)$ is simple if and only if w and all its derived words are presimple.*

In other words, if w does not correspond to a simple curve, then some derived word $w^{(n)}$ is not presimple and hence, when expressed in terms of some pair of generators $(\alpha, \beta) \in \underline{\Theta}$, contains one of the excluded blocks of Lemma 2.3 relative to the derived σ-sequences described below. If w represents a simple closed curve on M, then the the sequence of derived words terminates with a term $A^{(n)}$ or $B^{(n)}$ which is itself a simple curve appearing in $\underline{\Theta}$. If w is infinite non-periodic, the derivation process continues indefinitely and w is the χ-sequence of a leaf of one of the geodesic laminations referred to in Theorem 2(v).

The proof of Theorem 2 now proceeds as follows.

Step 1. We show that if $\sigma(\gamma)$ contains any excluded blocks then $e.h.(\gamma) > \nu_0$ ($e.h.(\gamma) > 3/2$ in the case of G_3).

Step 2. We associate *derived fundamental regions* to derived generating sets, set up σ-sequences relative to these new regions, and repeat the arguments of step 1.

Step 2 becomes rather involved and we shall not dwell long on it here. Although it is crucial if one wishes to complete the proof of Theorem 1 for G_3 by symbolic dynamics, it so happens in the case of G_5 that if $\gamma \notin \underline{\Theta}_C$ then $e.h.(\gamma) > \nu_0$, so that consideration of derived sequences is only needed to study the finer behaviour of essential heights moving along branches of the tree. The idea of using derived fundamental regions comes from Cohn [4]. A derived fundamental region $R^{(n)}$ for H_5 is an ideal decagon bounded by vertical sides at $\pm \omega$, which is a fundamental region for H_5 in which opposite sides are identified, in the same pattern as in R, by the derived generating set $A^{(n)}, B^{(n)}, C^{(n)}, D^{(n)}$. The first derived region for H_5 under the derivation Δ is shown in Figure 4. This corresponds to generators $A^{(1)} = B^{-1}A$, $B^{(1)} = B$, $C^{(1)} = B^{-1}C$, $D^{(1)} = B^{-1}D$. Any derived region $R^{(n)}$ is devided in two by a vertical geodesic meeting **R** at the central cusp between the two vertical sides. This line contains $\text{Ax}\, A^{(n)} \cap \text{Ax}\, B^{(n)}$ and $J^{(n)}$ is defined to be rotation by π about this point. The map T (and this is crucial) remains invariant under this process , i.e. $(B^{(n)})^{-1} J^{(n)} A^{(n)} = B^{-1}JA$. Derived σ-symbols are defined as before relative to the two halves of $R^{(n)}$.

We now return to step 1, in which the use of symbolic dynamics is fundamental. The idea is judiciously to choose a lift of an excluded block β so that any geodesic whose σ-sequence contains β is forced to have end points rather far apart on **R**. As a simple example, in the case of G_3, if $\beta = 2^3 \subset \sigma(\gamma)$, then γ can clearly be placed so that $|\gamma^+ - \gamma^-| > 3$, so that $ht(\gamma) > 3/2$.

This example is unfortunately deceptively easy. In general, the absence of excluded blocks can only be detected by examining arbitrarily long sequences in $\sigma(\gamma)$, so that one expects the height variation between curves which are and are not excluded to be arbitrarily small. (It is this fact on which the assertion of Theorem 2(iii) is based.) This is where symbolic dynamics comes into its own: one can check relative positions of points on **R** without having to do too many messy calculations.

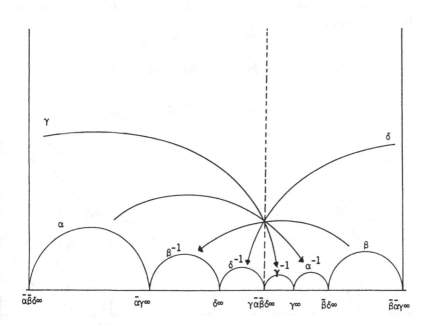

Figure 4 — A fundamental region for H after derivation

For definiteness we shall discuss the case G_5; the case of G_3 is by and large similar. Most of the blocks we need to consider contain the σ-sequence $2^3 3^2$. These are lifted as high in **H** as possible: by inspection one finds that this is to a position in which the imaginary axis I is placed between the two segments of type 2. We symbolise this by $2I23^2$.

Next, we make an important observation about the behaviour of the map T. The portion $\mathcal{A}$ of the tesselation $T(S)$ lying beneath the geodesic with the end points $A^{-1}(\infty)$, $A^{-1}(0)$ is mapped exactly to the portion $\mathcal{B}$ lying beneath the geodesic joining $B^{-1}(0)$, $B^{-1}(\infty)$. In particular T translates symbolic sequences in $\mathcal{A}$ into symbolic sequences in $\mathcal{B}$. Thus any σ-sequence $I^{-1}23X_1X_2\ldots$, $X_i \in \{2,3\}$ starting on I pointing into the left half plane, maps under T to the sequence $I32X_1X_2\ldots$ starting on I pointing into the right half plane. Since such sequences have well defined endpoints on **R**, we have described the action of T on the interval $[A^{-1}(\infty), A^{-1}(0)]$ in a very simple symbolic way. Further, these symbolic sequences can be used to compare relative positions on **R**, for example,

$$I32323\ldots < I32322\ldots,$$

as is clear by looking at the arrangement of copies of S in $\mathcal{B}$.

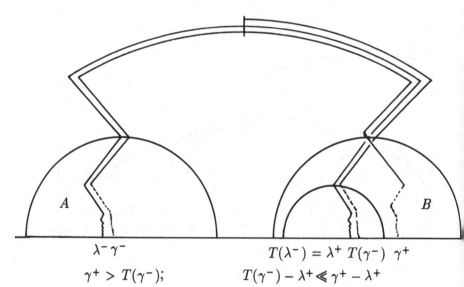

$$\lambda^- \gamma^- \qquad\qquad T(\lambda^-) = \lambda^+ \; T(\gamma^-) \; \gamma^+$$
$$\gamma^+ > T(\gamma^-); \qquad\qquad T(\gamma^-) - \lambda^+ \ll \gamma^+ - \lambda^+$$

Figure 5 — Comparison for the block $\beta = 232^2 32$

The idea is to compare all curves to the *limiting curve* $\lambda = \ldots 23232I23^2 2323\ldots$. This has χ-sequence $A^\infty B^{-1} A^\infty$, and has the limiting height ν_0. (In the classical case G_3 it has height $3/2$). It lies higher than all the curves

$$\ldots A^n B^{-1} A^n B^{-1} \ldots = \ldots 2^3 3^2 (23)^{n-1} 2I23^2 (23)^{n-1} \ldots, \quad n \geq 1,$$

in $\underline{\Theta}_C$ and, in the G_5 case, lower than any others. Since $\lambda^- = I^{-1}3232\ldots$ (note the sybols 2, 3 are interchanged when the orientation is reversed), the comments above on the action of T show that $T(\lambda^-) = I233232\ldots = \lambda^+$. Alternatively, this may be verified directly on observing that $\lambda^- = A^-$ and $\lambda^+ = B^{-1}(A^+)$.

We shall illustrate how the comparison works with one particular example, illustrated in Figure 5. Consider the excluded block $\beta = 232^2 32$. This is lifted to position $\tilde{\beta} = 232I232$ for comparison to λ. Let γ denote any geodesic such that $\sigma(\gamma)$ contains $\tilde{\beta}$. We wish to show that $\gamma^+ - \gamma^- > 2\nu_0$. Now $\gamma^+ = I232\ldots$, $\gamma^- = I^{-1}323\ldots$ and so $T(\gamma^-) = I23^2\ldots$. Hence comparison of symbols gives $\gamma^+ > T(\gamma^-)$. If T were parabolic, we would have

$$T(\gamma^-) - \gamma^- = T(\lambda^-) - \lambda^- = \lambda^+ - \lambda^- = 2\nu_0$$

and we would be done. The game now is to show that the error caused by the distortion in T is smaller than the excess of γ^+ over $T(\gamma^-)$. By direct computation,

$$T'(A^{-1}(\infty)) = 1.3707\ldots \quad \text{and} \quad T'(A^{-1}(0)) = 0.7294\ldots.$$

Thus the interval $[A^{-1}(\infty), A^{-1}(0)]$ contains the right hand boundary of the isometric circle of T and the distortion of T on this interval can be controlled. Moreover, as is most easily seen by inspecting Figure 5, the points λ^+ and $T(\gamma^-)$ lie in an interval bounded by a pentagon on the fourth level of the tesselation $T(S)$, while γ^+ is separated from $T(\gamma^-)$ by at least one side of a pentagon on the third level. Since one expects pentagons on higher levels to diminish in size, one hopes that these two effects balance. This indeed turns out to be the case. The estimates of size of small pentagons are carried out by repeated use of the identity

$$|g(\xi) - g(\eta)| = |g'(\xi)g'(\eta)|^{1}|\xi - \eta|, \quad \xi, \eta \in \mathbf{R}, \quad g \in G_5,$$

all of which is detailed in [14].

Let us simply comment here that one can see that the hyperbolicity of T is causing the slight variation between the limiting values ν_0 and ν_1 in Theorem 2.

It is clear that the methods described above would allow an attack on the other Hecke groups G_q. It follows as above from [6] that one should restrict to curves in $\underline{\Theta}_q$ where the tree $\underline{\Theta}_q$ is generated by the pair of generators A_q, B_q whose axes pass as close as possible to the centre of the tesselating q-gons as described in [6]. The map T_q is still hyperbolic and one presumes that its distortion will still cause splitting of bounding values analogous to ν_0, ν_1, however no detailed calculations have been made.

References

[1] A. Beardon and J. Lehner and M. Scheingorn, 'Closed geodesics on a Riemann surface with application to the Markoff spectrum', *Trans. A.M.S.* **295** (1986) 635–647.

[2] J. W. S. Cassels, *An introduction to Diophantine approximation*, (Cambridge University Press, 1957).

[3] M. Cohen and W. Metzler and A. Zimmerman, 'What does a basis of F(a,b) look like?', *Math Ann.* **257** (1981) 435–445.

[4] H. Cohn, 'Approach to Markoff's minimal forms through modular functions', *Ann. Math.* **61** (1955) 1–12.

[5] A. Haas, 'Diophantine approximation on hyperbolic Riemann surfaces', *Acta. Math.* **156** (1986) 33–82.

[6] A. Haas and C. Series, 'The Hurwitz constant and Diophantine approximation on Hecke groups', *J. London Math. Soc.* **34** (1986) 219–234.

[7] A. Leutbecher, 'Über die Heckeschen Gruppen $G(\lambda)$', *Abh. Math. Seminar Universität Hamburg,* **31** (1976) 199–205.

[8] F. F. Markoff, 'Sur les formes binairs indefinies I', *Math.Ann.* '**15** (1879) 381-406, II, **17** (1880) 379–400.

[9] R. Rankin, 'Diophantine approximation and horocyclic groups', *Canad. Math. J.* **9** (1957) 277–290.

[10] D. Rosen, 'A class of continued fractions associated with certain properly discontinuos groups', *Duke Math. J.* **21** 549–564.

[11] C. Series, 'The geometry of Markoff numbers' *Math. Intelligencer* **7** (1985) 20–29.

[12] C. Series,'The modular group and continued fractions', *J. London Math. Soc.* **31** (1985) 69–80.

[13] C. Series, 'Geometrical Markov coding of geodesics on surfaces of constant negative curvature', *Erg. Th. and Dyn. Sys.* **6** (1986) 601–625.

[14] C. Series, 'The Markoff spectrum in the Hecke group G_5', *Proc. London Math. Soc.* **57** (1988) 151–181.

This article is an abbrieviated version of [14].

5

On badly approximable
numbers, Schmidt games
and bounded orbits of flows

S. G. Dani
Tata Institute of Fundamental Research, Bombay, India.

§1. Introduction

A well-known result on badly approximable numbers motivates certain interesting results about abundance of bounded geodesics, bounded orbits of flows etc.. It is the purpose of this article to discuss these results explaining the underlying ideas.

We begin by recalling that $\alpha \in \mathbf{R}$, namely a real number is said to be *badly approximable* if there exists a $\delta > 0$ such that

$$|\alpha - p/q| > \delta/q^2$$

for all integers p, q and $q \neq 0$. The set of badly approximable numbers, which we shall denote by B, is a familiar object in the theory of Diophantine approximation (cf.[2],[19] etc.). It is a set of Lebesgue measure zero. Nevertheless it is still a large set in the sense that its Hausdorff dimension (with respect to the usual metric) is 1, which is the maximum possible for any subset of $\mathbf{R}$. This result, due to V. Jarník (cf.[18]) was strengthened by W. M. Schmidt who showed it to be a winning set for a certain game, which implies in particular, that given an open interval I and a sequence of functions $f_1, f_2, \ldots$ on I satisfying certain conditions, the set of α in I such that $f_i(\alpha) \in B$ for all i, is also of Hausdorff dimension 1. This holds for instance if each f_i is differentiable and the derivative is continuous and nowhere vanishing [16]. On the other hand it also holds if f_i are uniformly bi-Lipschitz maps, in the sense specified below.

Since we shall be dealing with many sets which are 'large' in a sense similar to the above, for brevity we introduce the following definition; specifically the condition is motivated by the latter form, which seems to be more convenient in higher dimensions.

We first recall that if X_1 and X_2 are metric spaces with metrics d_1 and d_2 respectively, a sequence of maps $f_i : X_1 \to X_2$, $i = 1, 2, \ldots$ is said to be uniformly bi-Lipschitz if there exists a $C > 1$ such that

$$C^{-1} d_1(x, y) \le d_2\big(f_i(x), f_i(y)\big) \le C \, d_1(x, y)$$

for all $x, y \in X_1$ and $i = 1, 2, \ldots$; a $C > 1$ for which the relations hold is called a uniform bi-Lipschitz constant for the sequence.

Definition. *A subset S of a metric space X is said to be incompressible if for any nonempty open subset Ω of X and any sequence $\{f_i\}$ of uniformly bi-Lipschitz maps of Ω onto (possibly different) open subsets of X, the subset*

$$\bigcap_{i=1}^{\infty} f_i^{-1}(S)$$

has the same Hausdorff dimension as X.

The above mentioned result can now be restated as follows.

Theorem 1.1. *The set B of badly approximable numbers is an incompressible subset of $\mathbf{R}$ (in the usual metric).*

We shall next observe a certain geometric significance of the set B. Let $\mathbf{H} = \{z \in \mathbf{C} : \operatorname{Im} z > 0\}$ be the upper half plane equipped with the usual Poincaré metric (cf.[1],[20]). The geodesics in $\mathbf{H}$ are either vertical lines or semicircles with endpoints in $\mathbf{R}$. By a geodesic ray we shall mean a curve of the form $\{g(t) : t \ge 0\}$, where $\{g(t) : t \in \mathbf{R}\}$ is a geodesic. Thus any geodesic ray ρ in $\mathbf{H}$ either has a unique limit point in $\mathbf{R}$ or it is directed vertically upward; we call the unique limit point or respectively ∞, the endpoint of ρ and denote it by $e(\rho)$. We recall also that there is an action of the group $SL(2, \mathbf{R})$ on $\mathbf{H}$ as a group of isometries, where the action of $\begin{pmatrix} a & b \\ c & d \end{pmatrix}$ is given by assigning to $z \in \mathbf{H}$ the element $(az + b)(cz + d)^{-1}$. The restriction of the action to the subgroup $\Gamma = SL(2, \mathbf{Z})$, consisting of integral unimodular matrices, is properly discontinuous and the quotient $M = \mathbf{H}/\Gamma$ is a noncompact surface, known as the modular surface. The geometric significance of B alluded to above lies in the following.

Proposition 1.2. *Let ρ be a geodesic ray in* H. *Then the image of ρ in $M = $ H$/\Gamma$, under the natural quotient map, is a bounded (relatively compact) subset of* H$/\Gamma$ *if and only if $e(\rho) \in B$.*

This can be proved using a well-known fundamental domain for the action of Γ on H and a fairly straightforward computation. In §7 we shall indicate the argument in a more general set up.

The modular surface M has two singularities, namely the orbits of $i = \sqrt{-1}$ and $\omega = \sqrt[3]{1}$ in H. For simplicity, in the following discussion we shall leave these out of consideration. Now let p be a nonsingular point in M and let S_p denote the space of tangent vectors of unit norm to M at the point p. For any $u \in S_p$ let $\sigma(p,u) = \{\sigma_{p,u}(t) : t \geq 0\}$ be the geodesic ray in the direction of u starting at p. Let $z \in$ H be a preimage of p. Then for $u \in S_p$, $\sigma(p,u)$ is the image of a unique geodesic ray in H starting at z, which we shall denote by $\rho(p,u)$.It is easy to see that the assignment $u \rightarrow e(\rho(p,u))$ defines a smooth map of $S_p \setminus \{u_p\}$ on to R, where u_p is the unique vector in S_p such that $e(\rho(p,u_p)) = \infty$. In view of this, Theorem 1.1 and Proposition 1.2 imply the following.

Theorem 1.3. *Let p be a nonsingular point in M and let S_p etc. be as above. Then the set*

$$C = \{u \in S_p : \sigma(p,u) \text{ is bounded in } M\}$$

is an incompressible subset of S_p.

§2. Bounded geodesics and horocycles

An interesting point about Theorem 1.3 is that though we use arguments involving the boundary in its proof, the assertion itself is intrinsic to the surface M. This suggests the question whether a similar assertion is true for other surfaces and more generally for Riemannian manifolds of higher dimension. Indeed the following turns out to be true.

Theorem 2.1. *Let M be a complete noncompact Riemannian manifold such that all the sectional curvatures are bounded between two negative constants and the Riemannian volume is finite. Let $p \in M$ and S_p be the space of unit tangent vectors at p. Let C be the subset of S_p consisting of all the elements u such that* *the geodesic ray starting at p in the direction of u is a bounded subset of M. Then C is an incompressible subset of S_p.*

In §7 we shall sketch a proof of this in the case when the curvature is constant. The case of variable curvature involves various technical points and will be dealt with elsewhere.

Let M be a manifold as in Theorem 2.1 and consider the geodesic flow corresponding to M (cf.[11]), defined on the unit tangent bundle, viz.

$$S(M) = \{(p, u) : p \in M, \ u \in S_p\}$$

equipped with the usual Riemannian metric. Theorem 2.1 then readily implies the following result on the dynamics of the flow.

Corollary 2.2. *Let the notation be as above and let C be the subset of $S(M)$ consisting of all elements (p, u) whose orbit under the geodesic flow is a bounded subset of $S(M)$. Then C is an incompressible subset of $S(M)$, with respect to the distance function induced by the Riemannian metric.*

We recall that a geodesic flow as above admits (one and hence a dense G_δ set of) points with dense orbits (cf.[11]). Also it is known to be ergodic, with respect to the Riemannian measure, when the curvature is constant (cf.[15]); I believe ergodicity holds for all M as above, though no reference is available since, in dealing with ergodicity in general authors assume M to be compact. When ergodicity holds, the set C as in the corollary is of measure zero, though it is large in the sense of being incompressible.

Corollary 2.2 for the geodesic flows is in complete contrast to what happens in the case of the horocycle flows associated to surfaces of constant negative curvature and finite area. In the latter case, by a classical result, due to G.A. Hedlund, there are a finite number (equal to the number of cusps of the surface) of embedded cylinders consisting of points with periodic orbits and all the remaining orbits are dense (cf.[10] for a stronger result). Thus the Hausdorff dimension of the set of points on bounded orbits is 2, though the dimension of the manifold of definition is 3.

The horocycle flow mentioned above corresponds to the action of the one-parameter subgroup

$$\left\{ \begin{pmatrix} 1 & t \\ 0 & 1 \end{pmatrix} : t \in \mathbf{R} \right\}$$

on $SL(2,\mathbf{R})/\Gamma$ where Γ is a lattice in $SL(2,\mathbf{R})$ (namely a discrete subgroup such that the quotient has finite invariant measure). A remark similar to the above also applies to orbits of the so called horospherical flows; the closures of orbits of these flows turn out to be closed orbits of possibly larger subgroups (cf.[6]). There is a conjecture due to M. S. Raghunathan that a similar assertion holds for actions of subgroups consisting of unipotent elements on homogeneous spaces G/Γ, where G is a Lie group and Γ is a lattice in G (cf.[4] for details).

Very recently it was also shown by G. A. Margulis [14] that if H is the subgroup of $SL(3,\mathbf{R})$ consisting of matrices which leave invariant the quadratic form $2x_1x_3 - x_2^2$ then any bounded orbit of the action of H on $SL(3,\mathbf{R})/SL(3,\mathbf{Z})$ is in fact a compact H-orbit; it may be worthwhile to note here that via this result, Margulis settled a conjecture of A. Oppenheim on the solvability of Diophantine inequalities of the form $|Q(x_1,\ldots,x_n)| < \epsilon$, for any $\epsilon > 0$, where Q is an indefinite quadratic form, asserting that this holds whenever $n \geq 3$ and Q is not a scalar multiple of a form with rational coefficients.

In [14] Margulis also asserts that his method yields a proof of Raghunathan's conjecture for bounded orbits of unipotent one-parameter subgroups of $SL(3,\mathbf{R})$ acting on $SL(3,\mathbf{R})/SL(3,\mathbf{Z})$. This would imply, in particular, that the set of points on bounded orbits of such a flow has Hausdorff dimension at most 6 as against dimension 8 of the space.

§3. Anosov flows

In the light of the discussion in the last section one may ask which flows on noncompact spaces have large sets of points with bounded orbits, while at the same time the typical orbit being dense. Since the geodesic flows as above are Anosov flows, more specifically one may ask the following

Question. Let $\{\phi_t\}$ be an Anosov flow on a noncompact Riemannian manifold M with finite Riemannian volume. Let C be the set of all the points x in M such that the orbit $\{\phi_t(x) : t \in \mathbf{R}\}$ is bounded (relatively compact). Then is C an incompressible subset of M?

We do not know the answer to the question in that general

form. However, apart from the geodesic flows, a well-known class of Anosov flows arising from certain flows on homogeneous spaces of Lie groups, which we describe below, admit an affirmative answer to the question.

Let G be a connected Lie group and Γ be a lattice in G. Suppose that G admits a one-parameter subgroup $A = \{a_t : t \in \mathbf{R}\}$ and a compact subgroup K satisfying the following conditions:

(a) $ka_t = a_t k$ for all $k \in K$ and $t \in \mathbf{R}$ and

(b) if U and U^- are the subgroups defined by

$$U = \{g \in G : a_t g a_{-t} \to e \text{ as } t \to \infty\}$$

and

$$U^- = \{g \in G : a_{-t} g a_t \to e \text{ as } t \to \infty\}$$

where e is the identity element in G, then $UKAU^-$ contains an open neighbourhood of the identity.

Suppose that Γ contains no nontrivial elements of finite order, so that the infrahomogeneous space (double coset space) $M = K\backslash G/\Gamma$ is a manifold. Then it can be verified that the flow $\{\phi_t\}$ on M defined by

$$\phi_t(Kg\Gamma) = Ka_t g\Gamma$$

is an Anosov flow with respect to a natural Riemannian structure. We refer the reader to P. Tomter [21] for various details including conditions for the existence and other aspects of subgroups as above. We recall in particular that the Lie group G as above has to be of $\mathbf{R}$-rank 1 and that $\mathrm{Ad}(a_t)$, $t \in \mathbf{R}$ have to be semisimple (diagonalisable) as linear transformations.

This construction yields a large class of Anosov flows. In fact it was conjectured earlier that any Anosov flow on a compact Riemannian manifold is topologically conjugate to one arising as above. That however has turned out to be false (cf.[13]); the author is thankful to A.Verjovsky for pointing this out.

To answer the question affirmatively for the Anosov flows as above it is evidently enough to prove the following.

Theorem 3.1. *Let G be a connected Lie group of $\mathbf{R}$-rank 1 and let Γ be a lattice in G. Let G/Γ be equipped with a metric induced by a right-invariant Riemannian metric on G. Let $\{g_t\}$ be a one-parameter subgroup of G consisting of semisimple elements (that* is, $\mathrm{Ad}(g_t)$ *is semisimple for all* t). *Let* C *be the subset of* G/Γ *consisting of all* $x\Gamma$, *where* $x \in G$, *such that the orbit* $\{g_t x\Gamma : t \in \mathbf{R}\}$ *is a bounded subset of* G/Γ. *Then* C *is an incompressible subset of* G/Γ.

We note that in the case when $G = SL(2,\mathbf{R})$, $\Gamma = SL(2,\mathbf{Z})$ and

$$ g_t = \begin{pmatrix} e^t & 0 \\ 0 & e^{-t} \end{pmatrix} \quad \text{for all } t \in \mathbf{R}, $$

the assertion of the theorem corresponds to Theorem 1.1 via Proposition 1.2 and certain simple facts. For $G = SO(n,1)$, the orthogonal group associated to a quadratic form of signature $(n,1)$, and Γ without any nontrivial elements of finite order, the theorem is essentially equivalent to the constant curvature case of Theorem 2.1.

Theorem 3.1 as stated above is slightly stronger than Theorem 5.1 of [8]. In §8 we shall indicate briefly how the strengthening can be achieved.

At this stage we caution the reader that while we were motivated to consider Anosov flows as a first step of generalisation from the geodesic flows, the Anosov property is certainly not a necessary condition for the existence of an incompressible set of points with relatively compact orbits. In the next section we shall see that this holds for certain flows on $SL(n,\mathbf{R})/SL(n,\mathbf{Z})$, which are not Anosov flows, when $n \geq 3$.

§4. Flows on $SL(n,\mathbf{R})/SL(n,\mathbf{Z})$

Let $G = SL(n,\mathbf{R})$ and $\Gamma = SL(n,\mathbf{Z})$, where $n \geq 2$. Let $1 \leq p \leq n-1$ and let $\{D_p(t)\}$ be the one-parameter subgroup consisting of diagonal matrices

$$ D_p(t) = \mathrm{diag}(e^{-t}, \ldots, e^{-t}, e^{\lambda t}, \ldots, e^{\lambda t}), $$

there being p entries of e^{-t} and $(n-p)$ entries of $e^{\lambda t}$, where $\lambda = p/(n-p)$. We consider the flow induced by the action of $\{D_p(t)\}$ on G/Γ. Let $q = (n-p)$,

$$ P = \left\{ \begin{pmatrix} A & B \\ 0 & D \end{pmatrix} \right\}, $$

where A, B and D are $p \times p$, $p \times q$ and $q \times q$ matrices respectively and $(\det A)(\det B) = 1$ and

$$N = \left\{ \begin{pmatrix} I & 0 \\ L & I \end{pmatrix} \right\},$$

where L is any $q \times p$ matrix, 0 and I being the zero and identity matrices of appropriate sizes.

It is easy to see that for $x \in P$, $\{D_p(t)xD_p(-t) : t \geq 0\}$ is bounded in G. Consequently for any $g \in G$ and $x \in P$, $\{D_p(t)g\Gamma : t \geq 0\}$ is bounded in G/Γ, if and only if $\{D_p(t)xg\Gamma : t \geq 0\}$ is bounded. Any $g \in G$ can be expressed as $g = xy\sigma$, where $x \in P$, $y \in N$ and σ is a matrix leaving invariant the set $\{\pm e_i : i = 1, \ldots, n\}$, $e_1, \ldots, e_n$, being the standard basis of $\mathbf{R}^n$; that is, σ is a permutation matrix up to signs. Since any such σ is contained in Γ, the above observation implies that $\{D_p(t)y\Gamma : t \geq 0\}$ is bounded if and only if $\{D_p(t)g\Gamma : t \geq 0\}$ is bounded. In other words, we only need to consider orbits of elements of N.

A rectangular matrix L of order $q \times p$ can be thought of as a system of q linear forms in p variables; each row is treated as coefficients in a linear form. It turns out that the orbit of $\begin{pmatrix} I & 0 \\ L & 0 \end{pmatrix}\Gamma$ under $\{D_p(t) : t \geq 0\}$ being bounded is related to the corresponding system of linear forms being badly approximable in the sense of the theory of Diophantine approximation, recalled below. For any $x \in \mathbf{R}^p$ let $\|x\|$ denote the maximum of the absolute values of the coordinates of x. Also for any $t \in \mathbf{R}$ let $\rho(t)$ denote the distance of t from the nearest integer. Let $\{\ell_1, \ldots, \ell_q\}$ be a system of q linear forms in p variables. It is said to be *badly approximable* if there exists a $\delta > 0$ such that for any nonzero integral vector x in $\mathbf{R}^p$ we have

$$\max_{1 \leq i \leq q} \rho(\ell_i(x)) > \delta \|x\|^{-p/q}$$

It may be noted that when $p = q = 1$ the condition above corresponds precisely to the unique coefficient being a badly approximable number.

In [3], using the well-known Mahler criterion we deduced the following.

Proposition 4.1. *Let L be a $q \times p$ matrix and let $\ell_1, \ldots, \ell_q$ be* the corresponding system of linear forms. Let

$$g = \begin{pmatrix} I & 0 \\ L & I \end{pmatrix}.$$

Then $\{D_p(t)g\Gamma : t \geq 0\}$ is bounded in G/Γ if and only if the system $\{\ell_1, \ldots, \ell_q\}$ is badly approximable.

In [18] Schmidt studied the set of systems of badly approximable linear forms, say q forms in p variables, and concluded that they form a winning set, in the sense of §5 below, in the space of all systems consisting of q linear forms in p variables; consequently they form an incompressible subset in the sense of the present paper. This, together with Proposition 4.1 imply the following

Theorem 4.2. *The set C of all $g\Gamma \in G/\Gamma$ such that the orbit $\{D_p(t)g\Gamma : t \in \mathbf{R}\}$ is bounded in G/Γ, is an incompressible subset of G/Γ.*

At this stage it is not clear to the author if similar results hold for flows induced by other one-parameter subgroups consisting of semisimple elements acting on $SL(n, \mathbf{R})/SL(n, \mathbf{Z})$, which, needless to say, would be of interest to know.

§5. The Schmidt game

We now describe a game introduced by W.M. Schmidt [17] which is at the heart of the results on incompressible sets described in the earlier sections.

The game involves two players, say $\mathcal{A}$ and $\mathcal{B}$, two numbers α and β in the open interval $(0, 1)$, a complete metric space X and a subset S of X. The procedure is as follows: $\mathcal{B}$ starts the game by picking a closed ball B_0 in X, (with arbitrary positive radius and arbitrary center). Then $\mathcal{A}$ chooses a closed ball A_1 contained in B_0 with radius α times the radius of B_0. Next $\mathcal{B}$ chooses a closed ball B_1 contained in A_1 with radius β times that of A_1 and so on. In general for $k \geq 1$ after $\mathcal{B}$ has chosen a closed ball B_{k-1}, $\mathcal{A}$ chooses a closed ball A_k contained in B_{k-1}, of radius α times that of B_{k-1}, and then $\mathcal{B}$ chooses a closed ball B_k contained in A_k and of radius β times the radius of A_k. Since X is a complete metric space, to each such sample procedure there corresponds a unique point of X, namely the point of intersection of the sequence $A_1 \supset A_2 \supset \ldots$, or equivalently of $B_0 \supset B_1 \supset \ldots$.

To win the game (against $\mathcal{B}$) $\mathcal{A}$ is required to force the point of intersection to be in S. If he does not succeed then $\mathcal{B}$ is considered the winner.

A subset S of X is said to be an (α, β)-*winning set* (for $\mathcal{A}$) if there is a strategy by which $\mathcal{A}$ can win; namely he can make his choices of A_k in such a way, that, no matter how $\mathcal{B}$ chooses B_k's during his turns, the point of intersection belongs to S. A set is said to be an α-*winning set*, where $\alpha \in (0,1)$, if it is an (α, β)-winning set for all $\beta \in (0,1)$.

It is intuitively clear that an (α, β)-winning set has to be 'large' in an appropriate sense. In fact it is not difficult to see that if $\beta \le 2 - \alpha^{-1}$ then X is the only (α, β)-winning set; in other words, in this case there is a strategy for $\mathcal{B}$ to force the point of intersection to be any pre-assigned point (cf.[17], Lemma 5). It turns out however that if $\beta > 2 - \alpha^{-1}$ then there could exist proper subsets which are (α, β)-winning sets. Schmidt showed in particular that the set of badly approximable numbers is an (α, β)-winning set for all α, β such that $\beta > 2 - \alpha^{-1}$ (cf.[17], Theorem 3, see also [18]). The following slightly strengthened version of another result of Schmidt (cf. [17],§11) shows that, all the same, such sets also have to be rather large in a certain sense.

Theorem 5.1. (*cf.* [8], *Proposition 3.1*) *Let X be a closed ball in $\mathbf{R}^n$ for some $n \ge 1$, equipped with the usual metric. Let $\alpha, \beta \in (0,1)$ and let S be an (α, β)-winning subset of X. Then for any nonempty open subset Ω of X the Hausdorff dimension of $S \cap \Omega$ is at least*

$$\log c_n \beta^{-n} / |\log \alpha\beta|$$

where c_n is independent of α and β. In particular if S is an α-winning set in X for some $\alpha \in (0,1)$ then $S \cap \Omega$ as above has Hausdorff dimension n.

We shall actually conclude that α-winning sets in $\mathbf{R}^n$ are incompressible. We first recall the following.

Proposition 5.2. (*cf. [17],§6*) *Let X be a complete metric space and let $\alpha \in (0,1)$. If $S_1, S_2 \ldots$ is a sequence of α-winning sets in X then so is $\bigcap_{i=1}^{\infty} S_i$.*

We also need the following observation which is straightforward to verify.

Proposition 5.3. *Let X and Y be complete metric spaces and let $\phi : X \to Y$ be a homeomorphism such that ϕ and ϕ^{-1} are both Lipschitz maps. Let a and b be the Lipschitz constants for ϕ and ϕ^{-1} respectively. Let $0 < \alpha < 1$ and $0 < \beta < (ab)^{-1}$. Let S be an (α, β)-winning set in Y. Then $\phi^{-1}(S)$ is an $(\alpha(ab)^{-1}, \beta ab)$-winning set in X.*

In [17] Schmidt considers the images of (α, β)-winning sets under what he calls 'local isometries' and proves a result which is much sharper than Proposition 5.3; in the above notation, for a local isometry ϕ we would have $\phi^{-1}(S)$ to be an $(\alpha c^{-1}, \beta c)$-winning set for *any* $c > 1$. In **R** it turns out that every continuously differentiable function with nowhere vanishing derivative is a local isometry. However in higher dimensions this does not hold unless the derivative is a scalar transformation. We therefore content ourselves, for the present, with Proposition 5.3. We now prove:

Corollary 5.4. *For any $\alpha \in (0, 1)$ any α-winning set S in $\mathbf{R}^n$, $n \geq 1$, is incompressible.*

Proof If $\alpha > 1/2$ then $S = \mathbf{R}^n$ and the assertion is obvious. Now let $\alpha \leq 1/2$ and let S be an α-winning set in $\mathbf{R}^n$. It is easy to see that if R is the closed subset of $\mathbf{R}^n$ bounded by a smooth hypersurface then the interior $\mathring{R}$ of R is an α-winning set in R and hence so is $S \cap \mathring{R}$. Now let Ω be a nonempty open subset of $\mathbf{R}^n$ and $\{f_i\}$ be a sequence of uniformly bi-Lipschitz maps of Ω onto open subsets of X with say $C > 1$ as a uniform bi-Lipschitz constant. Let X be a closed ball in Ω, with say radius $r > 0$. Using the above observation we can find, for all $i \geq 1$, a closed subset Y_i of $f_i(X)$ such that $S \cap Y_i$ is an α-winning set in Y_i and the Hausdorff distance between the boundaries of Y_i and $f_i(X)$ is less than $C^{-1}r/2$. Now let $Z_i = f_i^{-1}(Y_i)$ and $Z = \bigcap_{i=1}^{\infty} Z_i$. Then the latter condition ensures that Z contains a closed ball W of positive radius and the former together with Propositions 5.3 and 5.2 implies that $W \cap T$, where $T = \bigcap_{i=1}^{\infty} f_i^{-1}(S)$, is a $C^{-2}\alpha$-winning set in W. Hence by Proposition 5.1, $Z \cap T$ has Hausdorff dimension n, which proves the Corollary.

§6. Examples of winning sets in $\mathbf{R}^n$

We now describe certain winning sets in $\mathbf{R}^n$, needed in the proofs of Theorems 2.1. and 3.1. For a subset S of $\mathbf{R}^n$ the *thickness* $\tau(S)$ of S is defined to be

$$\tau(S) = \inf_V \sup_{x,y \in S} d(x - y, V),$$

where d denotes the usual distance function and the infimum is taken over all hyperplanes V in $\mathbf{R}^n$.

Theorem 6.1. *(cf.[8], Theorem 3.2) Let $\{S(j,t)\}$ be a family of subsets of $\mathbf{R}^n$ (doubly) indexed over $j \in \mathbf{N}$ and $t \in (0,1)$. Suppose that for any compact subset C of $\mathbf{R}^n$ and $\mu \in (0,1)$ there exist $M \geq 1$, $\sigma \in (0,1)$ and a sequence $\{\tau_j\}$ of positive numbers such that the following conditions are satisfied:*

a) if $j \in \mathbf{N}$ and $t \in (0,\sigma)$ are such that $S(j,t) \cap C$ is nonempty, then $\tau_j \leq M$ and $\tau(S(j,t)) \leq t\tau_j$ and

b) if $j,k \in \mathbf{N}$ and $t \in (0,\sigma)$ are such that $S(j,t) \cap C$ and $S(k,t) \cap C$ are both nonempty and $\mu\tau_j \leq \tau_k \leq \mu^{-1}\tau_j$ then either $j = k$ or $d(S(j,t),S(k,t)) \geq \sigma(\tau_j + \tau_k)$.

Then the set

$$W = \bigcup_{\delta>0} \left(\mathbf{R}^n \backslash \bigcup_{j=1}^{\infty} S(j,\delta)\right)$$

is an (α,β)-winning set for all $\alpha,\beta \in (0,1)$ such that $\beta > 2 - \alpha^{-1}$.

As a particular instance of this, we get the following.

Corollary 6.3. *Let $\{x_j\}$ be a sequence of distinct points in $\mathbf{R}^n$ and $\{r_j\}$ be a sequence of positive numbers such that $r_j \leq r$ for all j, for a suitable $r > 0$. Suppose that for any two distinct indices j and k*

$$d(x_j, x_k) \geq \sqrt{r_j r_k}.$$

Then the set

$$W = \bigcup_{\delta>0} \left(\mathbf{R}^n \backslash \bigcup_{j=1}^{\infty} B(x_j, \delta r_j)\right)$$

is an (α,β) -winning set for all $\alpha,\beta \in (0,1)$ such that $\beta > 2 - \alpha^{-1}$.

In particular if we let $n = 1$, $\{x_j\}$ the sequence of rational numbers (arranged in some order) and $r_j = 1/q^2$ if $x_j = p/q$, where $q \neq 0$ and p and q are coprime integers, then we get that the set of badly approximable numbers is an (α, β)-winning set for all $\alpha, \beta \in (0, 1)$ such that $\beta > 2 - \alpha^{-1}$.

6.3 Remark The condition in the corollary holds, in particular, in the following situation. Let us think of $\mathbf{R}^n$ as a horizontal hyperplane in $\mathbf{R}^{n+1}$, the extra direction being viewed as the vertical direction. Consider a sequence $\{B_j\}$ of mutually disjoint balls in $\mathbf{R}^{n+1}$ tangential to $\mathbf{R}^n$, lying 'over' the hyperplane. Now for $j \geq 1$ let $x_j \in \mathbf{R}^n$ be the point of tangency with B_j and let r_j be the radius of B_j. It is then easy to see that the condition in the corollary is satisfied for the sequences $\{x_j\}$ and $\{r_j\}$.

§7. Back to bounded geodesics

We shall now sketch a proof of Theorem 2.1 for manifolds, say M, of constant negative curvature. (cf.[7] for details). Such a manifold can be viewed as $\mathbf{H}^n/\Gamma$, where $\mathbf{H}^n$ is the hyperbolic n-space and Γ is a group of isometries acting properly discontinuously on $\mathbf{H}^n$ (cf.[22]). We view $\mathbf{H}^n$ as $\mathbf{R}^{n-1} \times (0, \infty)$ and call the second component the vertical coordinate. Since M is assumed to have finite volume, it has only finitely many ends (cf.[12]). Let us assume for simplicity that there is only one end; the general case can be dealt with using the same idea together with Proposition 5.2. Transforming the picture by an isometry if necessary, we can then arrange so that the Γ-action as above has a fundamental domain F whose (vertical) projection to $\mathbf{R}^{n-1}$ is relatively compact and the projection on the vertical line $(0, \infty)$ is contained in (a, ∞) for some $a > 0$. For $s > 0$ let H_s be the set of all points whose vertical component exceeds s. For all large s, H_s correspond to unfoldings in the covering of endpieces (complements of compact subsets) in M. It follows therefore that a geodesic ray $\rho = \{\rho_t : t \geq 0\}$ in $\mathbf{H}^n$ projects to a relatively compact subset of M if and only if it is wholly contained in the complement of $\Gamma(H_s)$ for some $s > 0$.

For each $\gamma \in \Gamma$ and $s > 0$ either $\gamma(H_s) = H_s$ or it is an open ball in $\mathbf{R}^n$ (in the euclidean metric) whose boundary is tangential to $\mathbf{R}^{n-1}$ at a point independent of s, usually denoted by $\gamma(\infty)$; further, the radius of the ball can be verified to have the form $c_\gamma s^{-1}$, where c_γ is independent of s. Also there exists $s_0 > 0$ such that for each fixed $s \geq s_0$, distinct elements of the family $\{\gamma(H_s) : \gamma \in \Gamma\}$ are mutually disjoint. For $\gamma \in \Gamma$ such that $\gamma(H_s)$

is an open ball and $t \in (0,1)$ let $S(\gamma, t)$ be the open ball in $\mathbf{R}^{n-1}$ with center at $\gamma(\infty)$ and radius $c_\gamma t s_0^{-1}$; namely the projection of $\gamma(H_{s_0 t^{-1}})$ on $\mathbf{R}^{n-1}$. Corollary 6.2 and Remark 6.3 then imply that the set

$$W = \bigcup_{\delta > 0} (\mathbf{R}^{n-1} \setminus \bigcup_{\gamma \in \Gamma} S(\gamma, \delta))$$

is an (α, β)-winning set for all $\alpha, \beta \in (0,1)$ such that $\beta > 2 - \alpha^{-1}$. In particular, by Corollary 5.4, it is incompressible.

Let $\rho = \{\rho_t : t \geq 0\}$ be a vertical geodesic ray pointing downward and let $e \in \mathbf{R}^{n-1}$ be its endpoint. Then ρ is wholly contained in the complement of $\Gamma(H_s)$ for some s (and hence for all $s' \geq s$) if and only if $e \in W$. Therefore by our earlier remark ρ projects to a bounded geodesic ray in M if and only if $e \in W$. We note however that any two geodesic rays in $\mathbf{H}^n$ with the same endpoint are asymptotic and consequently their images in M are either both bounded or both unbounded. Therefore we may conclude that for any geodesic ray ρ in $\mathbf{H}^n$ the image in M is bounded if and only if the endpoint of ρ belongs to W.

Now let $p \in M$ and $z \in \mathbf{H}^n$ be an element projecting to p. Let $\eta : S_p \longrightarrow \mathbf{R}^{n-1} \bigcup \{\infty\}$ be the map taking $u \in S_p$ to the endpoint of the geodesic ray starting at z in the direction of the (unique) unit tangent vector at z corresponding to u under the covering map. There is a unique element $u_\infty \in S_p$ such that $\eta(u_\infty) = \infty$ and η is smooth on the complement of $\{u_\infty\}$. Since W is incompressible this implies that $\eta^{-1}(W)$ is incompressible. This proves the theorem for the case at hand since $\eta^{-1}(W)$ is precisely the set described in the statement of the theorem.

§8. Comments on the proofs of other results

The proof of Theorem 2.1 in the variable curvature case proceeds along the same lines. However instead of the open balls $S(\gamma, t)$ as above we need to deal with horodiscs and use certain comparison theorems to show that the conditions of Theorem 6.1 are satisfied for appropriate sets. The technical details involved in this will be carried out elsewhere.

In the case of Theorem 3.1, the strategy involves difficulties of another kind. For a general group G as in the theorem there is no natural geometrical boundary to work with. In [8] this difficulty was overcome by employing the (abstract) Furstenberg boundary; this entails proving appropriate analogues of various simple geometrical facts and the results obtained in this respect would also be of independent interest. The natural analogue of $\mathbf{R}^{n-1}$, as the boundary in §7 is then a 2-step nilpotent Lie group N. We consider it partitioned into orbits of a certain Abelian Lie subgroup V and show that each orbit, viewed as a euclidean space, intersects the set corresponding to W as in §7, denoted in [8] by X, in a (α, β)-winning set for all α, β such that $\beta > 2 - \alpha^{-1}$. This readily implies that X is incompressible in the sense of the present paper. The proof can then be completed using considerations of asymptotic behaviour being dependent only on the endpoint (cf.[8], Lemma 2.1).

§9. Bounded orbits and simultaneous Diophantine approximation

In this section we point out a connection of the bounded orbits of the flows $\{D_p(t)\}$ as in §4 with a question of simultaneous Diophantine approximation. For this purpose we first note the following.

Theorem 9.1. *Let $1 \le p \le n-1$ and $q = n - p$. Let $x_1, \ldots, x_p$ and $y_1, \ldots, y_q$ be elements of $\mathbf{R}^n$. Suppose that*

i) for all $1 \le i \le p$ and $1 \le j \le q$, x_i is orthogonal to y_j and

ii) if $a_1, \ldots, a_p$ and $b_1, \ldots, b_q$ are real numbers such that some a_i and some b_j are nonzero then neither of $a_1 x_1 + a_2 x_2 + \ldots + a_p x_p$ or $b_1 y_1 + b_2 y_2 + \ldots + b_q y_q$ is an integral vector.

Then for any $\epsilon > 0$ there exists $\gamma \in SL(n, \mathbf{Z})$ such that

$$\|\gamma x_i\| < \epsilon \text{ for all } i = 1, \ldots, p$$

and

$$\|{}^t\gamma^{-1} y_j\| < \epsilon \text{ for all } j = 1, \ldots, q$$

This can be deduced from Theorem 4.1 of [5]. A version involving only the first set of inequalities was proved earlier by a direct method in [9].

Now let $x_1, \ldots, x_p$, $y_1, \ldots, y_q \in \mathbf{R}^n$ be vectors satisfying conditions i) and ii) of the theorem. For $\gamma \in SL(n, \mathbf{Z})$ let $\|\gamma\|$ denote the maximum of the absolute values of all its entries. Given $\epsilon > 0$, how small a value can we have for $\|\gamma\|$ with γ a solution of the inequalities in the theorem? Here is an answer covering one aspect of the question, involving $\{D_p(t)\}$ as in §4.

Theorem 9.2. *Let $x_1, \ldots, x_p$ and $y_1, \ldots, y_q$ be as above. Let $g \in SL(n, \mathbf{R})$ be such that $g(x_i) = \lambda_i e_i$ for $i = 1, 2, \ldots, p$ and ${}^t g^{-1} y_j = \lambda_{p+j} e_{p+j}$ for $j = 1, 2, \ldots, q$ where $\lambda_1, \lambda_2, \ldots, \lambda_n \in \mathbf{R}$ and $e_1, e_2, \ldots, e_n$ is the standard basis of $\mathbf{R}^n$. Let $\bar{g} = gSL(n, \mathbf{Z}) \in SL(n, \mathbf{R})/SL(n, \mathbf{Z})$. Then the following conditions are equivalent:*
  i) *$\{D_p(t)\bar{g} : t \geq 0\}$ is a bounded subset of $SL(n, \mathbf{R})/SL(n, \mathbf{Z})$*
  ii) *there exists a constant $C > 0$ such that for any $\epsilon > 0$ there exists $\gamma \in SL(n, \mathbf{Z})$ such that*

$$\|\gamma x_i\| < \epsilon \quad \text{for all} \quad i = 1, \ldots, p$$

$$\|{}^t\gamma^{-1} y_j\| < \epsilon^{p/q} \quad \text{for all} \quad j = 1, \ldots, q$$

$$\|\gamma\| < C\epsilon^{-p/q} \quad \text{and} \quad \|{}^t\gamma^{-1}\| < C\epsilon^{-1}$$

This can be proved by making minor modifications in the proof of Theorem 3.5 in [3].

It may be recalled that the set of $\bar{g}$ satisfying condition i) above is an incompressible set of zero measure.

§10. Miscellaneous comments and questions

Analogously to bounded orbits on noncompact spaces one can, in general, consider orbits which are bounded away from a particular point (namely, orbits avoiding a suitable neighbourhood of the point) looking, in particular, for their abundance or otherwise. Several participants at the conference raised this question; the author is thankful to them. In this respect, using Theorem 6.1 the author is able to deduce the following

Corollary 10.1 [1] . *Let A be a hyperbolic automorphism of the two-dimensional torus T^2. Let C be the set of points $x \in T^2$ such*

---

[1] In a forthcoming paper of the author to appear in *Ergodic Theory and Dynamical Systems*, Corollary 10.1 has been generalised to a large class of automorphisms of multidimensional tori.

that the closure of the orbit $\{A^j(x) : j \in \mathbf{Z}\}$ does not contain the identity element. Then C is an incompressible subset of T^2.

It would be interesting to know whether similar results hold in greater generality.

§11. References

[1] A. Beardon, *The Geometry of Discontinuous Groups*, (Springer Verlag, 1983).

[2] J.W.S. Cassels, *An Introduction to Diophantine Approximation* , (Cambridge University Press, 1957).

[3] S.G. Dani, 'Divergent trajectories of flows on homogeneous spaces and diophantine approximation', *J. reine und angew. Math.* **359** (1985) 55–89.

[4] S.G. Dani, 'Dynamics of flows on homogeneous spaces: A survey', Proceedings of "Colloquio de Systemas Dinamicos" held at Guanajuato, Mexico; *Soc. Mat. Mexicana* 1985.

[5] S.G. Dani, 'On orbits of unipotent flows on homogeneous spaces II', *Ergod. Th. & Dynam. Sys.* **6** (1986) 167–182.

[6] S.G. Dani, 'Orbits of horospherical flows', *Duke Math J.* **53** (1986) 177–188.

[7] S.G. Dani, 'Bounded geodesics on manifolds of constant negative curvature', Preprint .

[8] S.G. Dani, 'Bounded orbits of flows on homogeneous spaces', *Comment. Math. Helvetici* **61** (1986) 636–660.

[9] S.G. Dani and S.Raghavan, 'Orbits of euclidean frames under discrete linear groups', *Isr. J. Math.* **36** (1980) 300–320.

[10] S.G. Dani and J. Smillie, 'Uniform distribution of horocycle orbits for fuchsian groups', *Duke Math. J.* **51** (1984) 185–194.

[11] P. Eberlein, 'Geodesic flows on negatively curved manifolds I', *Ann.of Math.* **95** (1972) 492–510.

[12] H. Garland and M.S. Raghunathan, 'Fundamental domains for lattices in **R**-rank 1 semisimple Lie groups', *Ann of Math.* **92** (1970) 179–326.

[13] M. Handel and W. Thurston, 'Anosov flows on new three manifolds', *Invent. Math.* **59** (1980) 95–103.

[14] G.A. Margulis, 'Formes quadratiques indefinies et flots unipo-
tents sur l'espaces homogenes', *C.R. Acad. Sci, Paris, Ser.I*,
304 (1987) 249–253.

[15] F. Mautner, 'Geodesic flows on symmetric Riemann spaces',
Ann. of Math. **65** (1957) 416–431.

[16] W.M. Schmidt, 'On badly approximable numbers', *Mathematika* **12** (1965) 10–20.

[17] W.M. Schmidt, 'On badly approximable numbers and certain
games', *Trans. Amer. Math. Soc.* **123** (1966) 178–199.

[18] W.M. Schmidt, 'Badly approximable systems of linear forms',
J. Number Theory **1** (1969) 139–154.

[19] W.M. Schmidt, *Diophantine Approximation*, (Springer Verlag, 1980).

[20] A. Terras, *Harmonic Analysis on Symmetric Spaces and Applications I* , (Springer Verlag, 1985).

[21] P. Tomter, 'Anosov flows on infrahomogeneous spaces', *Proceedings of Symposia in Pure Mathematics*, Vol. XIV,
Amer.Math.Soc., 1970.

[22] J.A. Wolf, *Spaces of Constant Curvature*, (McGraw Hill,
1967).

6
Estimates for Fourier coefficients of cusp forms

S. Raghavan and R. Weissauer

Tata Institute of Fundamental Research and Universität Mannheim

§1. Introduction

The object of this paper is to show that improvements in estimates for Fourier coefficients of Siegel cusp forms yield improvements in estimates for the associated Satake parameters obtained usually from the theory of spherical functions [4].

It may be of interest to draw the readers' attention to a connection of the order of magnitude of Fourier coefficents of cusp forms with a phenomenon of uniform distribution and through that with ergodic theory and dynamical systems. The well-known Sato-Tate conjecture, for example, asserts that the angles θ_p for prime p, defined via the Ramanujan function $\tau(n)$ by $\cos\theta_p = \tau(p)/(2p^{11/2})$ are uniformly distributed in $[0,\pi]$ with respect to the measure $(2\sin^2\theta/\pi)d\theta$ and it admits to a generalisation to cusp forms in general; it may also be relevant to look for such formulations in higher dimensions.

§2. Estimation of Satake parameters

Let $f(Z) = \sum_{T>0} a(T)\exp(2\pi i\operatorname{tr}(TZ))$ for Z in the upper half-plane $\mathbf{H}_n$ of degree n be the Fourier expansion of a non-constant cusp form f of degree n and weight k for the modular group $\Gamma_n = Sp_n(\mathbf{Z})$. Further, let f be a Hecke eigenform, i.e. for every symplectic similitude matrix M in $\mathcal{M}_{2n}(\mathbf{Q})$

$$M\begin{pmatrix} 0 & \mathbf{1}_n \\ -\mathbf{1}_n & 0 \end{pmatrix}\,{}^tM = \ell(M)\begin{pmatrix} 0 & \mathbf{1}_n \\ -\mathbf{1}_n & 0 \end{pmatrix},$$

$$T(M) = \Gamma_n M\Gamma_n = \coprod_j \Gamma_n M_j, \qquad M_j = \begin{pmatrix} A_j & B_j \\ 0 & D_j \end{pmatrix},$$

where $\mathbf{0}$ is the (n, n) zero matrix, $\mathbf{1}_n$ is the n-rowed identity matrix and $^t P$ denotes the transpose of P, we have, for $\lambda(M) \in \mathbf{C}$,

$$f|_k T(M) := \sum_j f|_k M_j =$$

$$\sum_j f\left((A_j Z + B_j)D_j^{-1}\right) |D_j|^{-k} = \lambda(M)f$$

(here and in the sequel, the determinant of a square matrix D is denoted by $|D|$; we abbreviate $^t BAB$ as $A[B]$).

It is known that the Hecke operators $T(M)$ generate over $\mathbf{C}$ an algebra

$$\mathbf{H} = \underset{p}{\oplus}\mathbf{H}_p,$$

the tensor product (extended over all primes p) of the local Hecke algebra $\mathbf{H}_p$ which is isomorphic (through the Satake homomorphism Q – see [1], p.261) to the ring $\mathbf{C}\left[X_0^{\pm 1}, X_1^{\pm 1}, \ldots, X_n^{\pm 1}\right]^W$ of invariants under the Weyl group W in the polynomial ring over $\mathbf{C}$ in $X_0, X_0^{-1}, X_1, X_1^{-1}, \ldots X_n, X_n^{-1}$. Each system of eigenvalues $\lambda(M)$ defines a character of $\mathbf{H}_p$, mapping $X_0, X_1, \ldots, X_n$ respectively to the Satake p-parameters $\alpha_0, \alpha_1, \ldots, \alpha_n$ associated with the eigenform f.

For the Fourier coefficients $a(T)$ of any cusp form f of weight k, the estimate $a(T) = O\left(|T|^{k/2}\right)$ is well-known. However, let us make the following

Assumption.

$$|a(T)|\,|T|^{-k/2} \le C \prod_{1 \le j \le n} |T_j|^{(x_j - 1)/2}$$

for a constant $C > 0$ where T_j denotes the principal (j,j) submatrix of T and $x_1, \ldots, x_n \in \mathbf{R}$.

We assume that the Satake parameters $\alpha_1, \ldots, \alpha_n$ which are defined only up to a substitution under the Weyl group W, already satisfy the conditions $|\alpha_1| \ge |\alpha_2| \ge \ldots \ge |\alpha_n| \ge 1$. Moreover, let us define numbers $\beta_1 \ge \beta_2 \ge \ldots \ge \beta_n \ge 1$ such that $(\beta_0, \beta_1, \ldots, \beta_n)$ is in the W-orbit of $(\gamma_0, \gamma_1, \ldots, \gamma_n)$ with

$$\gamma_j = p^{-(x_j + \ldots + x_n)} \qquad (1 \le j \le n)$$

and

$$\gamma_0 = p^{n(2k+n+1)/4-(x_1+2x_2+\dots+nx_n)/2}.$$

Under the **Assumption** above, we now prove the following

Proposition. *For any $\epsilon > 0$, there exists a constant $c = c(\epsilon)$ independent of p such that*

$$|\alpha_1 \dots \alpha_n| \leq c \cdot p^\epsilon \beta_1 \dots \beta_n \qquad (1 \leq j \leq n).$$

Proof. Without loss of generality, we may suppose that in the **Assumption** stated, equality holds for some T. We may, for example, take the least value C for which inequality holds for all T and work instead with a suitable sequence of matrices T' for which

$$C(T') := |a(T')||T'|^{-k/2} \prod_{1 \leq j \leq n} |(T')_j|^{(1-x_j)/2}$$

and $\lim_{T'} C(T') = C$.

Since f is an eigenform for the Hecke operator corresponding to $\Gamma_n M \Gamma_n$, we have

$$\lambda(M)a(T) = \sum_j |D_j|^{-k} a(D_j T A_j^{-1}) \exp\left(2\pi i \operatorname{tr}(T A_j^{-1} B_j)\right) \quad (1)$$

where, for the left cosets $\Gamma_n M_j$, we choose

$$M_j = \begin{pmatrix} A_j & B_j \\ 0 & D_j \end{pmatrix} \quad \text{with} \quad A_j = \begin{pmatrix} p^{k_1} & & \star \\ & \ddots & \\ 0 & & p^{k_n} \end{pmatrix}$$

and

$$A_j{}^t D_j = p^{k_0} \mathbb{I}_n$$

and the integers $k_i = k_i(M_j)$ uniquely determined. Hence

$$|\lambda(M)||a(T)| \leq \sum_j |D_j|^{-k} |a(D_j T A_j^{-1})|.$$

In view of our **Assumption**, we see that

$$|\lambda(M)| \le \sum_j |D_j|^{-k} |D_j A_j^{-1}|^{k/2} \prod_{i=1}^{n} |(D_j)_i (A_j^{-1})_i|^{(x_i-1)/2}$$

$$= \sum_j p^{-n k_0 k/2} \prod_{i=1}^{n} p^{(x_i-1)(i k_0/2 - k_1(M_j) - \cdots - k_i(M_j))}$$

$$= \sum_j \gamma_0^{-k_0} \prod_{i=1}^{n} \left(\frac{\gamma_i}{p^i} \right)^{k_i(M_j)} |\det A_j|^{n+1}$$

$$= \mu(M)$$

where, by ([1], p.253), μ is the character of H_p mapping $X_0, X_1, \ldots, X_n$ to $\beta_0, \beta_1, \ldots, \beta_n$ respectively.

From [3] we know that, for $T(M) = \mathcal{T}_{j,n-j}(p^2)$ as defined therein,

$$Q\left(\mathcal{T}_{j,n-j}(p^2)\right) = X_0^{-2} X_1 \ldots X_n \sum_{l=0}^{j} c_l(j, n-j) R_l(X_1, \ldots X_n) \quad (2)$$

where

$$R_l(X_1, \ldots X_n) := \sum_{\substack{\epsilon_i = 0, \pm 1 \\ |\epsilon_1| + \cdots + |\epsilon_n| = l}} X_1^{\epsilon_1} \ldots X_n^{\epsilon_n}, \qquad (0 \le l \le n),$$

and $c_l(j, n-j) := p^{l(2n+1-l)} c(j-l, n-j)$ for $0 \le l \le j$ are positive integers defined by the conditions

$$c(2j+1, m) = (p^{m+1} - 1) c(2j, m+1)$$

$$c(2j, m) = \sum_{l=0}^{j} (-1)^l \binom{2j+m}{j-l} p^{4j^2 + 4jm + 2j + (m^2+m)/2 - 2ml - l^2} \gamma(l, m)$$

where

$$\gamma(0, m) = 1$$

$$\gamma(l, m) = \left(\frac{p^{2m+2l} - 1}{p^{2l} - 1} + p^m \right) \prod_{t=1}^{l-1} \frac{p^{2m+2t} - 1}{p^{2t} - 1} \qquad (l \ge 1).$$

From these formulae, it can be checked that

$$c(j,m) = \binom{j+m}{[j/2]} p^{j^2+2jm+j+(m^2+m)/2}\,(1+O(1/p))$$

and as a result,

$$c_l(j,n-j) = \binom{n-l}{[(j-l)/2]} p^{(n^2+2nj+n-j^2+j)/2}\,(1+O(1/p)).\tag{3}$$

for $0 \le l \le j$, where $[x]$ stands for the greatest integer $\le x$. From (3) we can see immediately that the ratios $c_l(j,n-j)/c_m(j,n-j)$ are bounded from above by a constant independent of p. Then applying the inequality $|\lambda(M)| \le \mu(M)$ above to $p^{-1}T_{j,n-j}(p^2)$ in place of M, we obtain by ([1], p.257) that

$$\left| R_j(\alpha_1,\ldots,\alpha_n) + \sum_{l<j} \frac{c_l(j,n-j)}{c_j(j,n-j)} R_l(\alpha_1,\ldots,\alpha_n) \right|$$

$$\le R_j(\beta_1,\ldots,\beta_n) + \sum_{l<j} \frac{c_l(j,n-j)}{c_j(j,n-j)} R_l(\beta_1,\ldots,\beta_n).\tag{4}$$

Observing that for $l < j$, $R_l(\beta_1,\ldots,\beta_n)/R_j(\beta_1,\ldots,\beta_n)$ are bounded from above by a constant independent of p and using the induction hypothesis

$$|R_l(\alpha_1,\ldots,\alpha_n)| \le C_l R_l(\beta_1,\ldots,\beta_n) \quad \text{for} \quad l < j$$

for a constant C_j, we deduce from (4) that

$$|R_j(\alpha_1,\ldots,\alpha_n)| \le C_j R_j(\beta_1,\ldots,\beta_n)\tag{5}$$

for a constant C_j and hence (5) holds for all $j \le n$.

Let us now assume the Proposition proved for $j-1$ in lieu of j. Fixing a $\delta > 0$, there exists a unique integer r with $j \le r \le n$ such that

$$|\alpha_s| \ge |\alpha_j| p^{-\delta(s-j)} \quad (j \le s \le r)$$

and

$$|\alpha_{r+1}| \le |\alpha_j| p^{-\delta(r+1-j)} \quad \text{if} \quad r < n.$$

Applying (5) with $j = r$, we get, with constants C'_r, C''_r,

$$|\alpha_1 \ldots \alpha_r| \leq C'_r R_r(\alpha_1, \ldots, \alpha_n)(1 + p^{-\delta})$$
$$\leq C''_r (1 + p^{-\delta}) R_r(\beta_1, \ldots, \beta_n),$$

i.e.

$$|\alpha_1 \ldots \alpha_{j-1}||\alpha_j|^{r+1-j} \leq C''_r (1 + p^{-\delta}) R_r(\beta_1, \ldots, \beta_n) p^{\delta(r-j)(r-j+1)/2}.$$

Hence, using induction, we see that

$$|\alpha_1 \ldots \alpha_j| \leq D_r (1 + p^{-\delta}) |\alpha_1 \ldots \alpha_{j-1}|^{1-1/(r+1-j)}$$
$$\times R_r(\beta_1, \ldots, \beta_n)^{1/(r+1-j)} p^{\delta(r-j)/2}$$
$$\leq D'_r (1 + p^{-\delta}) p^{\delta(r-j)/2} p^{\delta}(\beta_1 \ldots \beta_{j-1})^{1-1/(r+1-j)}$$
$$\times R_r(\beta_1, \ldots, \beta_n)^{1/(r+1-j)}$$
$$\leq D''_r (1 + p^{-\delta}) p^{\delta(r-j+2)/2} (\beta_1 \ldots \beta_{j-1})^{1-1/(r+1-j)}$$
$$\times (\beta_1 \ldots \beta_r)^{1/(r+1-j)}$$
$$\leq D''_r (1 + p^{-\delta}) p^{\delta(r-j+2)/2} \beta_1 \ldots \beta_r.$$

The proposition now follows, on choosing $\delta = 2\epsilon/n$ and a sufficiently large constant C_ϵ. ∎

Applications.

1) For the Fourier coefficients $a(T)$ of a cusp form f of degree 2 and weight k, Kitaoka [2] has obtained the estimate $a(T) = O(|T|^{k/2-1/4+\epsilon})$. The parameters x_1, x_2 in our **Assumption** above corresponding to this estimate may be taken to be $x_1 = 1$, $x_2 = 1/2 + 2\epsilon$ and then $\beta_1 = p^{3/2+2\epsilon}$, $\beta_2 = p^{1/2+2\epsilon}$. From the Proposition, we now obtain, for any $\epsilon > 0$,

$$|\alpha_1| \leq Cp^{3/2+\epsilon}, \qquad |\alpha_1\alpha_2| \leq Cp^{2+\epsilon}$$

with some constant C.

2) In the case of a cusp form of degree n and weight k the Rankin-Selberg method yields for the Fourier coefficients $a(T)$, the estimate

$$a(T) = O(|T|^{k/2+\epsilon-x/2})$$

where $x := 1/(n+1+2[n/2]+1/(n+1))$. Here we may take $x_1 = \cdots = x_{n-1} = 1$, $x_n = 1+2\epsilon - x$ and then $\beta_1 = p^{n-x+2\epsilon}$,

$\beta_2 = p^{n-1-x+2\epsilon}, \ldots, \beta_n = p^{1-x+2\epsilon}$. The Proposition gives us now the bound

$$|\alpha_1 \ldots \alpha_j| \le C' p^{nj - j(j-1)/2 - jx + \epsilon}$$

with a constant $C' = C'(\epsilon)$.

Remark. *The arguments above can be extended to cover the case of congruence subgroups of Γ_n as well.*

§3. Modified Rankin-Selberg method

It has always been of interest to obtain good estimates for the Fourier coefficients of cusp forms, in connection with asymptotic formulas for number-theoretic functions. By estimating generalized Kloosterman sums, Kitaoka [2], as mentioned earlier, has derived for the Fourier coefficients $a(T)$ of a cusp form f of degree 2 and weight k, the bound $a(T) = O(|T|^{k/2 - 1/4 + \epsilon})$, which is the sharpest so far in this situation. By applying the Rankin-Selberg method to f along with the general Eisenstein series, we will show that

$$a(T) = O(|T|^{k/2 - 1/4 - 3/38 + \epsilon})$$

for any $\epsilon > 0$, provided that $|T|$ goes to infinity, with the 'minimum' of T kept fixed.

We follow the notation used in [7]. Let

$$G(\mathbf{R}) = Sp_2(\mathbf{R})$$
$$= \left\{ M = \begin{pmatrix} A & B \\ C & D \end{pmatrix} \ \middle| \ A, B, C, D \in \mathcal{M}_2(\mathbf{R}), \right.$$
$$\left. A^t D - B^t C = \mathbf{1}_2, \quad A^t B = B^t A, \quad C^t D = D^t C \right\},$$

$$\Gamma_2 = Sp_2(\mathbf{Z}),$$
$$N(\mathbf{R}) = \left\{ M = \begin{pmatrix} \mathbf{1}_2 & * \\ 0 & \mathbf{1}_2 \end{pmatrix} \in G(\mathbf{R}) \right\},$$
$$A(\mathbf{R}) = \left\{ \begin{pmatrix} A & 0 \\ 0 & {}^t A^{-1} \end{pmatrix} \in G(\mathbf{R}) \ \middle| \ A = \begin{pmatrix} a_1 & 0 \\ 0 & a_2 \end{pmatrix}, \quad a_1, a_2 > 0 \right\},$$
$$U(\mathbf{R}) = \left\{ \begin{pmatrix} U & 0 \\ 0 & {}^t U^{-1} \end{pmatrix} \in G(\mathbf{R}) \ \middle| \ U = \begin{pmatrix} 1 & 0 \\ * & 1 \end{pmatrix} \right\},$$

$$B(\mathbf{R}) = N(\mathbf{R})A(\mathbf{R})U(\mathbf{R})$$

$$K(\mathbf{R}) = \left\{ \begin{pmatrix} C & -S \\ S & C \end{pmatrix} \in G(\mathbf{R}) \,\middle|\, C + iS \quad \text{unitary} \right\}$$

and

$$P(\mathbf{R}) = GL_2(\mathbf{R})N(\mathbf{R})$$

$$= \left\{ \begin{pmatrix} * & * \\ \mathbf{0} & * \end{pmatrix} \in G(\mathbf{R}) \right\}$$

where $\mathbf{0}$ is the 2×2 zero matrix and where $GL_2(\mathbf{R})$ is always assumed to be imbedded in $G(\mathbf{R})$ via

$$A \mapsto \begin{pmatrix} A & \mathbf{0} \\ \mathbf{0} & {}^t A^{-1} \end{pmatrix}.$$

Every g in $G(\mathbf{R})$ has the Iwasawa decomposition $g = n_g u_g a_g k_g$ with $n_g \in N(\mathbf{R})$, $u_g \in U(\mathbf{R})$, $a_g \in A(\mathbf{R})$, $k_g \in K(\mathbf{R})$ and $b_g := n_g u_g a_g \in B(\mathbf{R})$. For g in $G(\mathbf{R})$ again, we have the unique representation $g = g_Z \cdot k_g$ with

$$g_Z = \begin{pmatrix} \mathbf{1}_2 & X \\ \mathbf{0} & \mathbf{1}_2 \end{pmatrix} \begin{pmatrix} Y^{1/2} & \mathbf{0} \\ \mathbf{0} & Y^{-1/2} \end{pmatrix} \quad \text{and} \quad Z = X + iY,$$

the point in $\mathbf{H}_2 \simeq G(\mathbf{R})/K(\mathbf{R})$ corresponding to g; here $Y^{1/2}$ is a positive square root of the positive definite matrix $Y := \operatorname{Im}(Z)$.

For $g \in G(\mathbf{R})$ and $\mathbf{s} = (s_1, s_2) \in \mathbf{C}^2$, the general Eisenstein series $E(g, \mathbf{s})$ is defined by the series

$$E(g, \mathbf{s}) = \sum_{\gamma \in B \cap \Gamma_2 \backslash \Gamma_2} \prod_{j=1}^{2} (a_j(\gamma g))^{2s_j + j} \qquad (6)$$

which converges absolutely and uniformly on compact subsets of $\mathbf{C}^2$ for which $\operatorname{Re}(s_2 - s_1) > 1/2$ and $\operatorname{Re}(s_1) > 1/2$; here $a_j(\gamma g)$ for $j = 1, 2$ are just the components of $a_{\gamma g}$ in the above decomposition of γg. We introduce the more convenient variables ρ, s by setting $\rho = 2(s_2 - s_1)$, $s = 2s_1$ in terms of which the conditions of absolute convergence of (6) become $\operatorname{Re}(s) > 1$, $\operatorname{Re}(\rho) > 1$. We shall freely use in the sequel the variables ρ and s in lieu of s_1, s_2. It is known

[7] that $E(g, \mathbf{s})$ admits a meromorphic continuation to the whole of $\mathbf{C}_2$ and further

$$\xi\left(\frac{1+s}{2}\right)\xi\left(\frac{1+\rho}{2}\right)\xi\left(\frac{1+\rho+s}{2}\right)\xi\left(\frac{1+\rho+2s}{2}\right)$$

$$\times E\left(g, (\tfrac{s}{2}, \tfrac{s+\rho}{2})\right) \qquad (7)$$

is holomorphic and symmetric in s_1 and s_2 and moreover invariant under $s_1 \mapsto -s_1$, $s_2 \mapsto -s_2$; here $\xi(z) := z(z-1)\pi^{-z}\Gamma(z)\zeta(2z)$ and ζ denotes the Riemann zeta function. In terms of the Selberg zeta function (for GL_2) namely,

$$\zeta\left(g, (s_1 + \tfrac{3}{4}, s_2 + \tfrac{3}{4})\right) = \sum_{\sigma \in B \cap GL_2(\mathbf{Z}) \backslash GL_2(\mathbf{Z})} \prod_{j=1}^{2} a_j(\sigma g)^{2s_j + j}$$

for $g \in G(\mathbf{R})$ and $\operatorname{Re}(s_j) > 1$, $\quad(j = 1, 2)$, we see that

$$E(g, \mathbf{s}) = \sum_{\gamma \in P \cap \Gamma_2 \backslash \Gamma_2} \zeta\left(u_{\gamma g} a_{\gamma g}, (s_1 + \tfrac{3}{4}, s_2 + \tfrac{3}{4})\right). \qquad (8)$$

Now, for $g = n_g u_g a_g k_g = g_Z k_g \in G(\mathbf{R})$ with the corresponding point Z in $\mathbf{H}_2$, it is clear that

$$(u_g a_g)\,{}^t(u_g a_g) = \begin{pmatrix} Y & \mathbf{0} \\ \mathbf{0} & Y^{-1} \end{pmatrix}$$

where $Y = \operatorname{Im}(Z)$. Thus $a_1^2(g) = a_1^2(u_g a_g) = Y_1$, the leading diagonal element of Y and $a_1^2(g)a_2^2(g) = |Y|$; more generally, for $\sigma \in GL_2(\mathbf{R})$, $a_1^2(\sigma u_g a_g) = (Y[{}^t\sigma])_1$ and $a_1^2(\sigma u_g a_g)a_2^2(\sigma u_g a_g) = |Y[{}^t\sigma]| = |Y|$. Hence the function u_ρ defined for $g \in G(\mathbf{R})$ and $\operatorname{Re}(\rho) > 1$ by the series

$$\sum_{\sigma \in B \cap GL_2(\mathbf{Z}) \backslash GL_2(\mathbf{Z})} a_1(\sigma u_g a_g)^{-(\rho+1)/2} a_2(\sigma u_g a_g)^{(\rho+1)/2}$$

is indeed a function of $Y = \operatorname{Im}(Z)$ where Z in $\mathbf{H}_2$ corresponds to g. This function, which may thus be legitimately denoted by $u_\rho(Y)$, is a Größencharacter in the sense of Maaß [5]; $u_\rho(Y[U]) = u_\rho(Y)$

for every U in $GL_2(\mathbf{Z})$ and $u_\rho(\lambda Y) = u_\rho(Y)$ for $\lambda > 0$. Actually, if we associate to Y, the point $z = x + iy$ in the complex upper halfplane by means of

$$W = \frac{1}{\sqrt{|Y|}} Y = \begin{pmatrix} y^{-1} & -x/y \\ -x/y & (x^2 + y^2)/y \end{pmatrix},$$

then $u_\rho(Y)$ turns out to be just

$$\sum_\sigma \left(W[^t\sigma] \right)_1^{-(\rho+1)/2}$$

which is essentially the Eisenstein series

$$y^{(1+\rho)/2} \sum_{(c,d)=1} \text{abs}(cz+d)^{-(1+\rho)}.$$

It can also be checked immediately that

$$u_\rho(Y) = |Y|^{-(s/2+\rho/4+3/4)} \zeta\left(g, \left(s_1 + \tfrac{3}{4}, s_2 + \tfrac{3}{4}\right)\right) \qquad (9)$$

and as a result, we could well write $\zeta(Y, \mathbf{s})$ for $\zeta(g, \mathbf{s})$.

Let $\mathcal{P}$ denote the space of 2-rowed positive definite matrices

$$Y = \begin{pmatrix} y_{11} & y_{12} \\ y_{12} & y_{22} \end{pmatrix}$$

and dv, the invariant volume element $|Y|^{-3/2} dy_{11} dy_{12} dy_{22}$ in $\mathcal{P}$. Then, for $\alpha \in \mathbf{R}$, $\rho = 2(s_2 - s_1)$ with $\text{Re}(\rho) > 1$ and any $T \in \mathcal{P}$, we have the following

Lemma. For $\text{Re}(s_1) > 1/4 - \alpha$ and $\text{Re}(s_2) > 1/4 - \alpha$, we have

$$\int_\mathcal{P} e^{-\operatorname{tr}(TY)} |Y|^{(s_1+s_2+2\alpha)/2} u_\rho(Y)\, dv$$

$$= \sqrt{\pi}\,\Gamma(s_1 + \alpha - 1/4)\Gamma(s_2 + \alpha - 1/4) u_\rho(T^{-1})|T^{-1}|^{(s_1+s_2+2\alpha)/2}.$$

Proof. Since

$$a_1(\sigma u_g a_g)^{-(\rho+1)/2} a_2(\sigma u_g a_g)^{(\rho+1)/2)} = (Y[^t\sigma])_1^{-(\rho+1)/2} |Y|^{(\rho+1)/4}$$

for $g = g_Z \cdot k_g$ in $G(\mathbf{R})$ and $\sigma \in GL_2(\mathbf{R})$, the left hand side is just

$$\int_{\mathcal{P}} e^{-\operatorname{tr}(TY)}|Y|^{s_2+\alpha+1/4} \sum_{\sigma \in B \cap GL_2(\mathbf{Z}) \backslash GL_2(\mathbf{Z})} (Y[{}^t\sigma])_1^{-(\rho+1)/2} \, dv$$

$$= \sum_{\sigma} \int_{\mathcal{P}} e^{-\operatorname{tr}(TY)} (Y[{}^t\sigma])_1^{-(\rho+1)/2} |Y|^{s_2+\alpha+1/4} dv$$

$$= \sum_{\sigma} \int_{\mathcal{P}} e^{-\operatorname{tr}(T[\sigma^{-1}]Y)} Y_1^{-(\rho+1)/2} |Y|^{s_2+\alpha+1/4} dv$$

$$= \sqrt{\pi}\, \Gamma(s_1 + \alpha - 1/4)\Gamma(s_2 + \alpha - 1/4)$$
$$\times \sum_{\sigma} \left((T[\sigma^{-1}])^{-1}\right)_1^{-(\rho+1)/2} \left|(T[\sigma^{-1}])^{-1}\right|^{s_2+\alpha+1/4},$$

applying a formula of Maaß ([5], p.76). The lemma is now immediate in view of our earlier remarks on u_ρ. ∎

We now proceed to apply the Rankin-Selberg method of in-
tegrating the Γ_2-invariant function $|Y|^k|f(Z)|^2$ for a cusp form
f against the general Eisenstein series $E(g, \mathbf{s})$ over a fundamental domain $\Gamma_2 \backslash \mathbf{H}_2$ with respect to the invariant volume element
$dv := |Y|^{-3} dX \, dY$.

Proposition. *For any cusp form*

$$f(Z) = \sum_{T>0} a(T) \exp(2\pi i \operatorname{tr}(TZ))$$

of degree 2 and weight k and for $\operatorname{Re}(s) > 1$, $\operatorname{Re}(\rho) > 1$, we have

$$\int_{\Gamma_2 \backslash \mathbf{H}_2} |Y|^k |f(Z)|^2 E(g, (s_1, s_2)) \, dv$$

$$= \frac{\sqrt{\pi}\, \Gamma(s_1 + k - 1)\Gamma(s_2 + k - 1)}{2(4\pi)^{s_1+s_2+2k-3/2}} \sum_{\{T\}} \frac{|a(T)|^2}{\epsilon(T)|T|^k} \frac{u_\rho(T^{-1})|T|^{(3-\rho)/4}}{|T|^{s/2}}$$

*where $\{T\}$ runs over the $GL_2(\mathbf{Z})$-equivalence classes of positive
definite half-integral matrices T and $\epsilon(T)$ is the number of U in
$GL_2(\mathbf{Z})$ with $T[U] = T$.*

Proof. By (8), the left hand side of the asserted relation becomes (after the undoubtedly valid interchange of integration over $\Gamma_2 \backslash H_2$ and the summation of γ over $P \cap \Gamma_2 \backslash \Gamma_2$)

$$\sum_{\gamma \in P \cap \Gamma_2 \backslash \Gamma_2} \int_{\Gamma_2 \backslash H_2} |Y|^k |f(Z)|^2 \zeta \left(u_{\gamma g} a_{\gamma g}, (s_1 + \tfrac{3}{4}, s_2 + \tfrac{3}{4}) \right) dv$$

$$= \frac{1}{2} \int_{P \cap \Gamma_2 \backslash H_2} |Y|^k |f(Z)|^2 \zeta \left(u_g a_g, (s_1 + \tfrac{3}{4}, s_2 + \tfrac{3}{4}) \right) dv$$

$$= \frac{1}{2} \int_{P/GL_2(\mathbb{Z})} \left(\int_{X = {}^t X \bmod 1} |f(X + iY)|^2 \, dX \right) |Y|^{k-3} \zeta \left(Y, (s_1 + \tfrac{3}{4}, s_2 + \tfrac{3}{4}) \right) dY$$

$$= \frac{1}{2} \int_{P/GL_2(\mathbb{Z})} |Y|^{k-3} \left(\sum_{T>0} |a(T)|^2 \exp(-4\pi \operatorname{tr}(TY)) \right) \zeta \left(Y, (s_1 + \tfrac{3}{4}, s_2 + \tfrac{3}{4}) \right) dY$$

$$= \frac{1}{2} \sum_{T>0} |a(T)|^2 \int_{P/GL_2(\mathbb{Z})} \exp(-4\pi \operatorname{tr}(TY)) |Y|^{k-3} \zeta \left(Y, (s_1 + \tfrac{3}{4}, s_2 + \tfrac{3}{4}) \right) dY$$

$$= \frac{1}{2} \sum_{T} \frac{|a(T)|^2}{\epsilon(T)} \int_{P} \exp(-4\pi \operatorname{tr}(TY)) |Y|^{(s_1 + s_2 + 2k - 3/2)/2} u_\rho(Y) \, dv \quad \text{by (8)}$$

The proposition now follows on using the Lemma. ∎

Let us now define the series $D(s, \rho)$ for $\operatorname{Re}(s) > 1$ and $\operatorname{Re}(\rho) > 1$, by

$$D(s, \rho) = \zeta(1+s)\zeta(1+\rho)\zeta(1+\rho+s)\zeta(1+\rho+2s)$$
$$\times \sum_{\{T\}} \frac{|a(T)|^2}{\epsilon(T)|T|^k} \frac{u_\rho(T^{-1})|T|^{(3-\rho)/4}}{|2T|^{s/2}} \qquad (10)$$

$$= \frac{2^{1-s}}{\sqrt{\pi}} (4\pi)^{s+\rho/2+2k-3/2}$$
$$\times \zeta(1+s)\zeta(1+\rho)\zeta(1+\rho+s)\zeta(1+\rho+2s)$$
$$\times \{\Gamma(k+s/2-1)\Gamma(k+\rho/2+s/2-1)\}^{-1}$$
$$\times \int_{\Gamma_2 \backslash H_2} |Y|^k |f(Z)|^2 E\left(g, (s/2, (\rho+s)/2) \right) dv,$$

taking our Proposition into account. From the functional equation of $E(g,s)$ and the identity $\Gamma(z)\Gamma(1-z) = \pi/\sin\pi z$, we obtain

$$D(-s,-\rho) = E(s,\rho)G(s,\rho)D(s,\rho)$$

where

$$E(s,\rho) = \frac{4^s \sin(\pi s)\sin(\pi(\rho+s))\cos(\pi\rho/2)\cos(\pi(\rho/2+s))}{4\pi^6(4\pi)^{\rho+2s}\pi^{3\rho+4s}}$$

and

$$G(s,\rho) = \Gamma^2\left(\frac{1+s}{2}\right)\Gamma^2\left(\frac{1+\rho}{2}\right)\Gamma^2\left(\frac{1+\rho+s}{2}\right)$$

$$\times \Gamma^2\left(\frac{1+\rho+2s}{2}\right)\Gamma\left(k+\tfrac{s}{2}-1\right)\Gamma\left(2-k+\tfrac{s}{2}\right)$$

$$\times \Gamma\left(k+\tfrac{\rho+s}{2}-1\right)\Gamma\left(2-k+\tfrac{\rho+s}{2}\right).$$

Now, while the series $D(s,\rho)$ converges absolutely for $\mathrm{Re}(s) > 1$ and $\mathrm{Re}(\rho) > 1$, it can be seen that $D(s,-\rho)$ also converges absolutely for $\mathrm{Re}(s) > 1 + \mathrm{Re}(\rho)$, $(\mathrm{Re}(\rho) > 1$ fixed$)$, as a Dirichlet series in s by using the functional equation

$$u_{-\rho}(T^{-1}) = \frac{\xi((1+\rho)/2)}{\xi((1-\rho)/2)}u_\rho(T^{-1}).$$

By analytic continuation, (10) holds also with ρ replaced by $-\rho$, in the region $\mathrm{Re}(s) > 1 + \mathrm{Re}(\rho)$, $\mathrm{Re}(\rho) > 1$.

We are now in a position to apply a theorem of Landau [6] to the Dirichlet series

$$D(2s_1,\rho) = \sum_{n=1}^{\infty} d_n(\rho)n^{-s_1}, \quad D(2s_1,-\rho)$$

with the respective abscissa of absolute convergence $1/2$, $(1+\rho)/2$ for any fixed (real) $\rho > 1$. If

$$c_n(\rho) := \sum_{\substack{\{T\} \\ |2T|=n}} \frac{|a(T)|^2}{\epsilon(T)|T|^k}u_\rho(T^{-1})|T|^{(3-\rho)/4}, \quad (n \geq 1),$$

then clearly

$$c_n(\rho) = \frac{1}{\zeta(1+\rho)} \sum_{l_1,l_2,l_3 \geq 1} \frac{\mu(l_1)\mu(l_2)\mu(l_3)}{l_1^{1+\rho} l_2^{1+\rho} l_3} d_{n/l_1^2 l_2^2 l_3^2}(\rho)$$

where $\mu(\cdot)$ is the Möbius function. We use Landau's theorem in the following simple form:

Let $\phi(s_1) = \sum_{n=1}^{\infty} d_n n^{-s_1}$, $d_i \geq 0$ converge absolutely for $\mathrm{Re}(s_1) \geq \mu_1$ and have a meromorphic continuation to the whole of $\mathbf{C}$ such that $R(s_1)\phi(s_1)$ is entire and of finite order for some polynomial $R(s_1)$. Suppose further that there exist $\mu_2, \alpha_1, \ldots, \alpha_\nu \in \mathbf{R}$, $\beta_1 > 0, \ldots, \beta_\nu > 0$ and $q_1, \ldots, q_\lambda, p_1, \ldots, p_\lambda$ in $\mathbf{C}$ such that for $\mathrm{Re}(s_1) \geq \mu_2$, $\phi(s_1)$ satisfies the functional equation

$$\phi(-s_1) = \prod_{i=1}^{\nu} \Gamma(\alpha_i + \beta_i s_1) \left(\sum_{j=1}^{\lambda} \exp(q_j + p_j s_1) \right) \psi(s_1)$$

where $\psi(s_1) = \sum_{m=1}^{\infty} e_m m^{-s_1}$ converges absolutely for $\mathrm{Re}(s_1) \geq \mu_2$. Then if $|\mathrm{Im}(p_i)| \leq \pi(\beta_1 + \ldots + \beta_\nu)/2$ for $1 \leq i \leq \lambda$ and

$$x := \mu_2 \sum_{i=1}^{\nu} \beta_i - (\nu+1)/2 + \sum_{j=1}^{\nu} \alpha_j > 0, \quad \mu_1 + \mu_2 > 0,$$

we have

$$\sum_{n \leq N} d_n = B(N) + O(N^\eta),$$

where $B(N)$ is the sum of the residues of $\phi(s_1)N^{s_1}/s_1$ and

$$\eta := \frac{\mu_1 x - \mu_2}{1 + x}.$$

In our situation involving $D(2s_1, \rho)$ and $D(2s_1, -\rho)$ we have for any $\epsilon > 0$,

$$\mu_1 = \frac{1}{2} + \epsilon, \quad \mu_2 = \frac{1+\rho}{2} + \epsilon, \quad \sum_i \alpha_i = 5 - 3\rho, \quad \sum_i \beta_i = 12,$$

$$\nu = 10, \quad x = (11)/2 + 3\rho + \epsilon, \quad \eta = \frac{9 + 4\rho}{26 + 12\rho} + \epsilon.$$

Therefore, we obtain

$$d_N(\rho) = O(N^\eta)$$

and as a consequence

$$\dot{c}_N(\rho) = \frac{1}{\zeta(1+\rho)} \sum_{l_1, l_2, l_3 \geq 1} \frac{1}{l_1^{1+\rho} l_2^{1+\rho} l_3} O\left(\left(\frac{N}{l_1^4 l_2^2 l_3^2}\right)^\eta\right)$$

$$= O(N^\eta).$$

Finally taking

$$T = \begin{pmatrix} t_{11} & t_{12} \\ t_{12} & t_{22} \end{pmatrix} > 0$$

to have t_{22} as its 'minimum' without loss of generality and $|2T| = N$, we have

$$\frac{|a(T)|^2}{\epsilon(T)|T|^k} = O\left(\frac{|T|^\eta}{|T|^{(3-\rho)/4} u_\rho(T^{-1})}\right)$$

for every $\rho > 1$. Using the trivial estimate $u_\rho(Y) > y^{(1+\rho)/2}$, we have $u_\rho(T^{-1}) > \left(\sqrt{|T|}/t_{22}\right)^{(1+\rho)/2}$ and hence

$$|a(T)|^2 = O\left(\frac{|T|^{k+\eta}(\min T)^{\frac{1+\rho}{2}}}{|T|}\right)$$

$$= O\left(|T|^{k-\frac{17+8\rho}{26+12\rho}+\epsilon}(\min T)^{\frac{1+\rho}{2}}\right).$$

Letting ρ tend to 1, we obtain the estimate

$$|a(T)| = O\left((\min T)^{1/2} |T|^{k/2-1/4-3/38+\epsilon}\right)$$

as asserted at the beginning of this section. ∎

Acknowledgement. The first-named author is thankful to Professor U. Christian and Professor E. Freitag for their kind hospitality.

§4. References

[1] E. Freitag, *Siegelsche Modulfunktionen,* Grundlehren math. Wissen. 254, Springer Verlag, 1983

[2] Y. Kitaoka, 'Fourier coefficients of Siegel cusp forms of degree two', *Nagoya Math. J.* **93**(1984) 149–171

[3] A. Krieg, 'Das Vertauschungsgesetz zwischen Hecke-Operatoren und dem Siegelschen Φ-Operator', *Arch. Math.* **46** (1986) 323–329

[4] R. P. Langlands, *Euler Products.*Yale University Press, 1971

[5] H. Maaß, *Siegel's Modular Forms and Dirichlet Series.*Lecture Notes in Mathematics 216, Springer-Verlag, 1971

[6] M. Sato and T. Shintani, 'On zeta functions associated with prehomogeneous vector spaces'. *Ann. of Math.* **100** (1974) 131–170

[7] R. Weissauer, 'Eisensteinreihen vom Gewicht $n+1$ zur Siegelschen Modulgruppe n-ten Grades'. *Math. Ann.* **268** (1984) 357-377

7
The integral geometry of fractals

K. J. Falconer

University of Bristol, Bristol, UK

§1. Fractals

Fractals occur in a wide variety of branches of mathematics, for examples as attractors or Julia sets in dynamical systems, or as sets of well-approximable numbers in number theory. It is desirable to have a geometric theory of fractals and dimension that can be applied with advantage in such a variety of situations. In this article we survey one part of this theory that has been developed in recent years, but which perhaps is not as widely known as its potential applicability warrants. A wide ranging account of 'The Fractal Geometry of Nature' is given in Mandelbrot's book [10]. An account of measure theoretic aspects is given by Falconer [3]. Fundamental to the theory of fractals is the notion of dimension.

The most frequently used definition is that of Hausdorff dimension, which is defined in terms of Hausdorff measures. Let $E \subset \mathbf{R}^n$ be any set, and suppose $0 \leq s \leq n$. For $\delta > 0$ let

$$\mathcal{H}_\delta^s(E) = \inf \left\{ \sum_{i=1}^\infty |U_i|^s : E \subset \bigcup_{i=1}^\infty U_i \quad \text{and} \quad 0 < |U_i| < \delta \right\}.$$

Here $|\ \ |$ denotes the diameter of a set, so the infimum is over all coverings of E by sets of diameter less than δ. Then $\mathcal{H}_\delta^s(E)$ increases as $\delta \to 0$, so that we may define the s-dimensional Hausdorff outer-measure of E as

$$\mathcal{H}^s(E) = \lim_{\delta \to 0} \mathcal{H}_\delta^s(E)$$

It is easy to show that $\mathcal{H}^s$ is an outer measure and restricts to a measure on the Borel subsets of $\mathbf{R}^n$. For any set $E \subset \mathbf{R}^n$ there is a value of s at which $\mathcal{H}^s(E)$ jumps from ∞ to 0 as s increases. Thus we define the *Hausdorff dimension* of E

$$\dim E = \sup \{s : \mathcal{H}^s(E) = \infty\} = \inf \{s : \mathcal{H}^s(E) = 0\}.$$

Hausdorff measures generalise Lebesgue measures to fractional dimensions, and satisfy the scaling property

$$\mathcal{H}^s(\lambda E) = \lambda^s \mathcal{H}^s(E)$$

where λE is obtained by scaling E by a factor λ about the origin. From these definitions we can show that the dimension of a rectifiable curve is 1, of a disc is 2, of the middle-third Cantor set is $\log 2/\log 3$, of the von Koch 'snowflake' curve is $\log 4/\log 3$, and so on. Fractals, which may be thought of as sets with a fine structure, typically have non-integral dimension, or have Hausdorff dimension different from their topological dimension.

§2. Integral Geometry

Classical integral geometry applies to smooth, or at least rectifiable sets that are anything but of a fractal nature. A fundamental formula is due to Poincaré: if E and F are rectifiable curves in the Euclidean plane then

$$\int (\text{number of points in } E \cap \sigma(F)) \, d\sigma$$

$$= 4(\text{length of } E)(\text{length of } F)$$

where integration is with respect to the natural invariant measure on the set of rigid motions σ (thus $d\sigma = d\theta \, dx \, dy$ where (x, y) is the point of the plane containing E to which an end of F is translated, and θ is the angle of rotation). Poincaré's formula is easily verified, first for a pair of line segments, then by polygonal approximation to general E and F. A similar idea applies if smooth manifolds E and F in $\mathbf{R}^n$ are moved rigidly relatively to each other. If they intersect at all, then 'in general' they will intersect in a set of dimension $\max\{0, \dim E + \dim F - n\}$. More precisely, if $\dim E + \dim F - n > 0$ then $\dim(E \cap \sigma(F)) = \dim E + \dim F - n$ for a set of rigid motions σ of positive measure, and is 0 for almost all other σ.

The problem we consider here is whether such results hold if E and F are fractals and we use Hausdorff dimension: for which $E, F \subset \mathbf{R}^n$ do we have

$$\dim(E \cap \sigma(F)) \leq \max\{0, \dim E + \dim F - n\} \text{ 'in general' (A)}$$

and

$$\dim(E \cap \sigma(F)) \geq \dim E + \dim F - n \quad \text{'often'} \qquad \text{(B)}$$

as σ ranges over an appropriate group G of transformations (e.g. translations, congruences or similitudes)? Of course, 'in general' means for almost all σ, and 'often' means for a set of σ of positive measure, with respect to the natural invariant measure on G. We restrict our discussion of the main results to the case when E and F are 'reasonable' sets such as Borel or compact sets such as occur in practice — otherwise very few positive results are possible! Detailed proofs are not given here, except in a couple of cases to try to indicate the 'flavour' of the subject. Full details of some of this work may be found in the treatises of Mattila [13,15], Federer [6] and Kahane [8,9].

§3. Towards Inequality A

As far as inequality A is concerned, all the positive results that there are hold when G is the group of translations, and so hold automatically for the larger groups of congruences and similarities. The following special case leads to the most general result. If $x \in \mathbf{R}^n$ and $F \subset \mathbf{R}^n$, we write $x + F$ for the vector sum $\{x + f : f \in F\}$.

Lemma 1. *Let* $E, F \subset \mathbf{R}^2$ *with* F *a line segment. Let* $1 \leq \dim E \leq 2$. *Then*

$$\dim(E \cap (x + F)) \leq \dim E - 1 = \dim E + \dim F - 2$$

for almost all $x \in \mathbf{R}^2$.

Proof. Let $\{U_i\}$ be a collection of convex sets covering E with $0 < |U_i| < \delta$ and

$$\sum_i |U_i|^s \leq \mathcal{H}_\delta^s(E) + \epsilon. \qquad (1)$$

Define a non-negative function supported by $\bigcup_i U_i$ by

$$f(x) = \sum_i |U_i|^{s-2} \chi_{U_i}(x) \qquad (x \in \mathbf{R}^2)$$

where $\chi_{U_i}(x)$ is the characteristic function of U_i. By Fubini's theorem

$$\ell \int_{\mathbf{R}^2} f(x)\, dx = \int_{x \in \mathbf{R}^2} \left\{ \int_{y \in x + F} f(y)\, dy \right\} dx$$

where ℓ is the length of the line segment F. Since the area of a set of diameter d is at most $\frac{1}{4}\pi d^2$ by the isoperimetric inequality, this gives

$$\ell \sum_i |U_i|^{s-2} \frac{1}{4}\pi |U_i|^2 \geq \ell \int f(x)\,dx$$

$$\geq \int_{x \in \mathbf{R}^2} \left\{ \sum_i |U_i|^{s-2} |U_i \cap (x+F)| \right\} dx$$

or

$$\frac{1}{4}\pi l \sum |U_i|^s \geq \int_{x \in \mathbf{R}^2} \left\{ \sum |U_i \cap (x+F)|^{s-1} \right.$$

for some collection $U_i \cap (x+F)$ that covers $\left. E \cap (x+F) \right\} dx.$

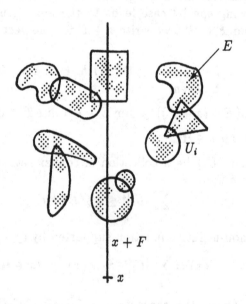

Figure 1

We can do this for any $\epsilon > 0$, so by (1)

$$\frac{1}{4}\pi\ell\mathcal{H}_\delta^s(E) \geq \int\limits_{x\in\mathbf{R}^2} \mathcal{H}_\delta^{s-1}(E\cap(x+F))\,dx;$$

letting $\delta \to 0$, we get

$$\frac{1}{4}\pi\ell\mathcal{H}^s(E) \geq \int\limits_{x\in\mathbf{R}^2} \mathcal{H}^{s-1}(E\cap(x+F))\,dx$$

and the result follows. ∎

Notice that a very similar argument works if F is any rectifiable curve. In the same way, if $E, F \subset \mathbf{R}^n$ and F is a d-dimensional flat (i.e. a subset of positive d–dimensional area of a translate of a d–dimensional subspace), then

$$\dim(E\cap(x+F)) \geq \max\{0, \dim E + \dim F - n\} \qquad (2)$$

for almost all $x \in \mathbf{R}^n$.

We now use Lemma 1 to obtain a more general result.

Theorem 1. *Let $E, F \subset \mathbf{R}^n$. Then*

$$\dim(E\cap(x+F)) \geq \max\{0, \dim(E\times F) - n\}$$

for almost all $x \in \mathbf{R}^n$.

Proof. We show this in the case of $E, F \subset \mathbf{R}$; the result follows for $n > 1$ in exactly the same way. Let L be the line $y = x$ in $\mathbf{R}^2$. Assuming that $\dim(E\times F) > 1$ we have that

$$\dim\big((E\times F)\cap((x,-x)+L)\big) < \dim(E\times F) - 1 \qquad (3)$$

for almost all $x \in \mathbf{R}$ by Lemma 1. (Here we make use of the redundancy of one of the coordinate dimensions as we translate L along itself.) But $(e, f) \in (E\times F)\cap((x,-x)+L)$ if and only if $e \in E$, $f \in F$, $e = x+y$ and $f = -x+y$ for some $y \in \mathbf{R}$, that is if $e = 2x + f$ or $e \in E\cap(2x+F)$. Projecting onto the x-axis we see that $(E\times F)\cap((x,-x)+L)$ and $E\cap(2x+F)$ are geometrically similar, so that (2) follows from (3) if $n = 1$. ∎

Although (2) is a very long way from (A), examples show that it is the best result we can hope to achieve. In general for any E, F

$$\dim(E \times F) \geq \dim E + \dim F \tag{4}$$

(see Eggleston [2] or Marstrand [11]). However in very many situations we do have equality in (4), in which case (2) reduces to (A). This happens, for example, if either E or F is a rectifiable curve, or under more general conditions given by Tricot [17] in terms of packing dimensions.

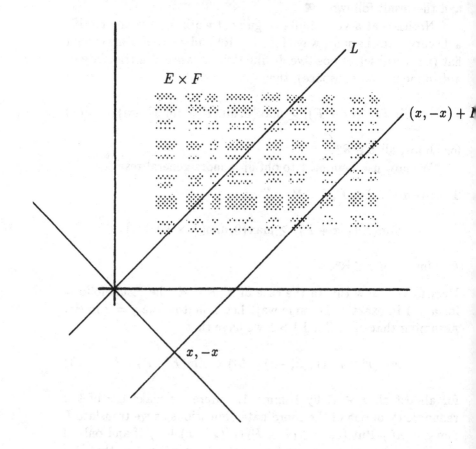

Figure 2

§4. Towards inequality B

Lower estimates for $\dim(E \cap \sigma(F))$ are generally harder to obtain. The following is known:

Theorem 2. *Let $E, F \subset \mathbf{R}^n$ be Borel sets, and G a group of transformations on $\mathbf{R}^n$. Then*

$$\dim(E \cap \sigma(F)) \geq \dim E + \dim F - n$$

for a set of motions $\sigma \in G$ of positive measure in the following cases:
(i) G is the group of similarities and E and F are arbitrary.
(ii) G is the group of congruences, E is arbitrary and F is a rectifiable curve or surface.
(iii) G is the group of congruences and E and F are arbitrary with either $\dim E > \frac{1}{2}(n+1)$ or $\dim F > \frac{1}{2}(n+1)$.

We omit the proofs, which are based on the potential theoretic characterisation of Hausdorff dimension: if E supports a mass distribution μ with $0 < \mu(E) < \infty$ such that

$$\iint\limits_{E\ E} |x - y|^{-s}\, d\mu(x)\, d\mu(y) < \infty$$

then $\dim E \geq s$, and, conversely, if $s < \dim E$ then there exists such a mass distribution on E with

$$\iint\limits_{E\ E} |x - y|^{-s}\, d\mu(x)\, d\mu(y) < \infty$$

(see Falconer [3] §6.2 for details of this equivalence). Part (i) is proved by Mattila [13], and Kahane [8]. Part (ii) may also be found in Mattila [14], though for F a straight line this essentially dates back to Marstrand's paper [12]. Mattila [14] modified his method to prove (iii) using a variant of a Fourier transform estimate of Falconer [4]. The transform method breaks down without the condition on $\dim E$ or $\dim F$, and it is not known whether inequality B holds if $n \geq 2$ and $\dim E$, $\dim F \leq \frac{1}{2}(n+1)$. However,

Mattila [13] constructs $E, F \subset \mathbf{R}$ both of dimension 1 and of the form

$$\bigcap_{i=1}^{\infty} \bigcup_{j=1}^{\infty} I_{i,j}$$

where the $I_{i,j}$ are intervals, such that $E \cap \sigma(F)$ contains at most one point for any translation σ, so that inequality B fails for congruences if $n = 1$.

all rotations 'equally likely'

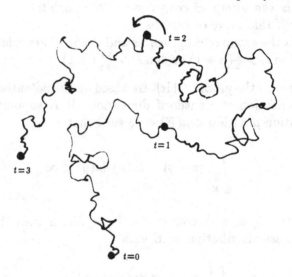

Figure 3

§5. Applications to Brownian Motion

Let $\omega(t)$ be a Brownian motion path in $\mathbf{R}^2$, so that the increments of the paths $\omega(t_2) - \omega(t_1), \ldots, \omega(t_{2m}) - \omega(t_{2m-1})$ are independent if $t_1 < t_2 < \ldots < t_{2m-1} < t_{2m}$, and $\omega(t + h) - \omega(t)$ has zero (vector) mean and variance h for all t. It has been known for a long time (Taylor [18]) that almost surely the path $\omega(t)$ has Hausdorff dimension 2. By Theorem 2 there exists a set of congruences σ of positive measure such that $\dim\big(\omega([0,1]) \cap \sigma(\omega([2,3]))\big) = 2$. By the isotropy of Brownian motion, given the point $\omega(2)$ and a realization of $\omega([2,3])$, this realization is equally likely to occur rotated about $\omega(2)$ in any direction. Moreover, given $\omega([0,1])$, the distribution of the position of $\omega(2)$ is absolutely continuous with respect to Lebesgue measure. It follows that there is a positive probability of $\omega([0,1]) \cap \omega([2,3])$ having dimension 2. By the self-similarity of the Brownian process, this same probability pertains in any time interval, so we conclude that the set of points on the path visited more than once has dimension 2 almost surely.

By repeating this argument, it follows that the set of points of multiplicity at least m has dimension 2 almost surely for every positive integer m. Similarly, Theorems 1 and 2 may be used to show that almost surely Brownian paths in $\mathbf{R}^3$ have double points but no triple points, and that Brownian motion paths in $\mathbf{R}^n$ ($n > 4$) have no multiple points. Results of this nature were first obtained by direct analysis of Brownian motion by Dvoretzky, Erdös and Kakutani [1], but it seems much more natural to deduce them from general results on fractals. This theme is developed by Kahane [9]; see Taylor [19] for a recent survey on such random processes.

§6. Sets with large intersection

In this section we consider a class of set where inequality A fails drastically, typically with $\dim(E \cap \sigma(F)) = \min\{\dim E, \dim F\}$. The sets we consider are everywhere uncountably dense, and are in general Borel but not compact.

For $E \subset \mathbf{R}^n$ and $0 \le s \le n$ let

$$\mathcal{M}^s(E) = \inf \left\{ \sum_{i=1}^{\infty} |U_i|^s : E \subset \bigcup_{i=1}^{\infty} U_i \right.$$

$$\left. \text{where the } U_i \text{ are any binary cubes} \right\}.$$

(A binary cube is of the form

$$\left[2^{-k} m_1, 2^{-k}(m_1 + 1) \right) \times \cdots \times \left[2^{-k} m_n, 2^{-k}(m_n + 1) \right)$$

where k is a non-negative integer and $m_1, \ldots, m_n$ are integers). Then $\mathcal{M}^s(E)$ is an outer measure of net type (but not a Borel measure) that is related to $\mathcal{H}^s(E)$. In particular $\mathcal{H}^s(E) = 0$ if, and only if, $\mathcal{M}^s(E) = 0$. A sequence of open subsets $\{F_r\}$ is called $\mathcal{M}^s$-*dense* in $\mathbf{R}^n$ if for every open $A \subset \mathbf{R}^n$

$$\lim_{r \to \infty} \mathcal{M}^s(F_r \cap A) = \mathcal{M}^s(A).$$

Typically the sets F_r will consist of collections of open intervals with spacing decreasing to 0 as $r \to \infty$ and with the lengths of the intervals decreasing not too rapidly compared with the spacing. We now define the class of sets

$$\mathcal{C}^s(\mathbf{R}^n) = \left\{ \limsup_{r \to \infty} \overline{F}_r : \{F_r\} \text{ is } \mathcal{M}^s \text{ dense in } \mathbf{R}^n \right\}.$$

These classes have the following properties.
(1) If $E \in \mathcal{C}^s$ then $\dim E \ge s$
(2) If $E \in \mathcal{C}^s$ and $F \supset E$ then $F \in \mathcal{C}^s$
(3) If $E_1, E_2, \ldots$ is a countable sequence with $E \in \mathcal{C}^s$ then $\bigcap_{i=1}^{\infty} E_i \in \mathcal{C}^s$. In particular $\dim \bigcap_{i=1}^{\infty} E_i \ge s$ by (1), so inequality A fails.

(4) Let $f : \mathbf{R}^n \to \mathbf{R}^n$ be any 'reasonable' smooth surjective function. If $E \in C^s$ then $f(E) \in C^s$ and $f^{-1}(E) \in C^s$.

Thus we could take $f_1, f_2, \ldots$ as any similarity mappings to get that $\dim \bigcap_{i=1}^{\infty} f_i(E) \geq s$ for and $E \in C^s$.

Many sets obtained as a 'lim sup' belong to C^s for an appropriate value of s. Such sets occur in Diophantine approximation in number theory, as the following examples show.

Let $\nu_r \searrow 0$ and $0 < s < 1$ and

$$E = \left\{ x \in \mathbf{R} : |x - \nu_r p| \leq \nu_r^{1/s} \right.$$

$$\left. \text{for some integer } p \text{ for infinitely many } r \right\}.$$

It may be shown that $E \in C^s$, so that $\dim E \geq s$, and so, for example $\dim \bigcap_{i=1}^{\infty} f_i(E) \geq s$ for any congruence transformations $f_1, f_2, \ldots$. In particular, we could take

$$E = \left\{ x \in \mathbf{R} : |x - p/r| \leq r^{-1/s} \right.$$

$$\left. \text{for some integer } p \text{ for infinitely many } r \right\}$$

These are the '$1/s$ well-approximable numbers' shown by Jarník [7] to have Hausdorff dimension $\min\{2s, 1\}$. However, $E \in C^{2s}$ if $0 < s < 1/2$, so taking $f_i(x) = x^k$, say, it follows from properties (1)–(4) that $\dim \bigcap_{i=1}^{\infty} f_i^{-1}(E) \geq 2s$ or

$$\dim \left\{ y \in \mathbf{R} : y^k \text{ is } 1/s \text{ well-approximable for all } k \right\} = 2s$$

Full details of the classes C^s and their properties may be found in Falconer [5]. Schmidt [16] also consider classes of 'α-winning sets' which exhibit properties like (1)–(4) above, but the dimensional properties that he derives appear to be rather coarser than for the classes C^s.

§7. References

[1] A. Dvoretzky, P. Erdös and S. Kakutani, 'Double points of paths of Brownian motion in n-space', *Acta Sci. Math.* **12** (1950) 75–81.

[2] H. G. Eggleston, 'The Besicovitch dimension of Cartesian product sets', *Proc. Camb. Phil. Soc.* **46** (1950) 383–386.

[3] K. J. Falconer, *The Geometry of Fractal Sets*, Cambridge University Press (1985).

[4] K. J. Falconer, 'On the Hausdorff dimension of distance sets', *Mathematika,* **32** (1985) 206–212.

[5] K. J. Falconer, 'Classes of sets with large intersection', *Mathematika,* **32** (1985) 191–205.

[6] H. Federer, *Geometric Measure Theory*, Springer, New York, (1969).

[7] V. Jarník, 'Über die simultanen diophantischen Approximationen', *Math. Zeit.,* **33** (1931) 505–543.

[8] J-P. Kahane, 'Sur la dimension des intersections', in *Aspects of Mathematics and its Applications*, ed. J.A. Barrow, 419–430, North Holland (1986).

[9] J-P. Kahane, 'Ensembles aleatoires et dimensions', in *Recent Progress in Fourier Analysis*, North Holland (1985).

[10] B. Mandelbrot, *The Fractal Geometry of Nature*, San Fransisco, W. H. Freeman (1982).

[11] J. M. Marstrand, 'The dimension of Cartesian product sets', *Proc. Camb. Phil. Soc.,* **50** (1954) 198–202.

[12] J. M. Marstrand, 'Some fundamental geometrical properties of plane sets of fractional dimensions', *Proc. Lond. Math. Soc.(3)* **4** (1954) 257–302.

[13] P. Mattila, 'Hausdorff dimension and capacities of intersections of sets in n-space', *Acta Math.,* **152** (1984) 77–105.

[14] P. Mattila, 'On the Hausdorff dimension and capacities of intersections', *Mathematika,* **32** (1985) 213–217.

[15] P. Matilla, *Lecture Notes on Geometric Measure Theory* , Departmento de Matematicás, Universidad de Extremadura (1986).

[16] W. M. Schmidt, 'Badly approximable numbers and certain games', *Trans. Amer. Math. Soc.,* **123** (1966) 178–199.

[17] C. Tricot, 'Two definitions of fractional dimension', *Math. Proc. Camb. Phil. Soc.* **91** (1982) 57–74.

[18] S. J. Taylor, 'The Hausdorff α-dimensional measure of Brownian paths in n-space', *Proc. Camb. Phil. Soc.*, **49** (1953) 31–39.

[19] S. J. Taylor, 'The measure theory of random fractals', *Math. Proc. Camb. Phil. Soc.*, **100** (1986) 383–406.

8
Geometry of algebraic continued fractals

J.Harrison [1]

University of California, Berkeley, USA

§1. Introduction

In mathematics, we are often faced with the problem of analyzing a sequence of real numbers x_n. The sequence may come from a purely mathematical setting or from a computer output. What is the asymptotic behaviour of x_n? Is it periodic, dense, uniformly distributed? What are the dynamics of the sequence? In this paper we discuss a method for 'graphing' x_n so that such properties become more apparent.

Computer scientists use *scattter plots* to achieve this goal with some sucess. Our method, invoking fractal geometry, improves on scatter plots.

Given a sequence of real numbers x_n there exists a Jordan curve Q in the plane called a **fractal** **plot***. The geometry of Q reflects analytical properties of x_n.*

We present in this paper the general algorithm for producing Q from x_n. The sequence $x_n = n\alpha \pmod 1$ where α is an irrational number is especially interesting and for such a sequence we call Q a *continued fractal*. We describe a dictionary for associating the geometry of Q with the analytical properties of x_n. For example, we prove that if α is quadratic then Q is asymptotically self-similar.

Scatter Plots. Formally, a sequence x_n is a function $f : \mathbf{Z} \to \mathbf{R}$ such that $f(n) = x_n$. The standard *graph* of f consists of the set of pairs $\{(n, f(n)) : n \in \mathbf{Z}\}$. This graph becomes more vivid by drawing a line segment between the points $(n, f(n))$ and $(n, 0)$. Since the graph will not fit on the screen for n large, the length of the line segments can be reduced by using a composition such as $n \to 1 - 1/n$. The net effect is a picture, called a *scatter*

---

[1] partially supported by USAF Grant No. F49620-87-C-0118

'bunch up' near the vertical line through $(1,0)$. You can see more detail by composing with $n \to 1 - 1/\sqrt{n}$, say, but there will always be some obscuring of the detail in the limit.

There are two orders inherent in a sequence - the 'time' order and its 'space' order. A good picture of a sequence should display these two orders well. Scatter plots make clear the time order, but not the space order except in simple examples. Properties such as density and uniform distribution are hidden.

The diagrams of Dekking and Mendes France Dekking and Mendes France [1] showed how to draw a curve in the plane for any sequence of real numbers x_n between 0 and 1. The curve consists of the line segments connecting

$$\sum_{n=1}^{k} e^{2\pi i x_n} \quad \text{and} \quad \sum_{n=1}^{k+1} e^{2\pi i x_n} \quad \text{for} \quad k \geq 1.$$

Again, the time order is displayed most clearly. Each x_n is represented by a segment L_n of length 1. The segments are attached to each other according to the index of the sequence: L_1 is attached to L_2 which is attached to L_3, etc. The location of x_n in $\mathbb{R}$ is represented by the slope of L_n. Thus the properties such as density and dynamics are not always easy to observe. However the picture is genuinely two dimensional and in some examples gives a significant improvement over scatter plots. It is not always bounded in the plane, but that could be changed by using weights $g(n)e^{2\pi i x_n}$ where $\sum g(n) < \infty$.

The Mathematician's sequence diagram. A mathematician wanting to visualize a sequence typically draws dots on the line and labels them $x_1, x_2, \ldots$. This method has obvious defects. For example, only a few points can be labeled. If only dots are drawn, then the sequence information is lost. If many dots are drawn, then all you may see is a 'blur'. The next technique improves on this:

Inverse scatter plots. For $1-1$ sequences we may define and graph

$$f^{-1} : f(\mathbb{Z}) \to \mathbb{Z}$$

This time we draw the line segments $(x,0)$ to $(x, f^{-1}(x))$ to emphasize the graph.

This time we draw the line segments $(x, 0)$ to $(x, f^{-1}(x))$ to emphasize the graph.

The picture can be made bounded by restricting the range of f to $[0, 1]$ and composing f^{-1} with an invertible $g : \mathbf{Z} \to [0, 1]$. Again, draw vertical line segments from $(x, 0)$ to $(x, g(f^{-1}(x)))$. The length L of the line segment over x clearly indicates its index n:

$$n = g^{-1}(L).$$

For a better picture, assume that $g(n)$ is monotone and $g(n) \to 0$ as $n \to \infty$.

Here, the space order of x_n is emphasized over its time order. This gives a better picture of x_n if one is interested in density or dynamics.

As with scatter plots, this graph has limitations. The topology of the graph is not simple: it is a line with many lines attached to it (or a "comb"). Asymptotically the graph begins to look like a lot of dots. The interplay between the time n and the position x_n becomes obscured.

We make one more modification to remedy these deficiencies. We make the comb into a continuous curve.

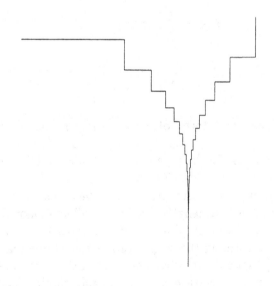

Figure 1 — The fractal plot for an orbit attracted to a fixed point

Fractal plots. Roughly, draw the inverse scatter plot for the first k terms of the sequence and then 'glue' into the base I the vertical line segments to make a continuous curve Q_k. That is, where the vertical line segment L meets I, make a cut in I. This cut will produce two points in the base where there was one. Attach one of these to the upper end point of L and the other to the lower endpoint of L. That is, where the vertical line segment L meets I, divide the interval between the preceding and the following vertical segments into two. Attach one of these to the upper end point of L and the other to the lower. This choice alternates to keep the figure within a bounded region — half of the segments L go up, half down. The horizontal segments from I will stay horizontal; they will be shifted up or down. See figure 1. We make this more precise:

The computer algorithm: Let x_n be a $1 - 1$ or a repeating sequence on the unit interval $I = [0, 1)$. For simplicity, assume $x_0 = 0$. Let $g : \mathbf{Z}^+ \to I$ be a monotone function, called the *weight* function. Let $0 \leq \theta \leq \pi/2$ and $k \in \mathbf{Z}^+$. We define an arc Q_k in the plane depending on the sequence $x_0, \ldots, x_k$, the weight g and the angle θ. Define two normalising constants:

$$g(0) = \sum_{n=1}^{k} (-1)^n g(n)$$

$$m_k = 1 + \sum_{n=0}^{k} g(n).$$

Order the first $k+1$ terms of the sequence x_k from left to right in I: $x_{n_0} = 0, x_{n_1}, \ldots, x_{n_k}$. Let $I_{n_i} = (x_{n_i}, x_{n_i+1})$ denote the complementary intervals. In the plane, draw a horizontal line segment with left endpoint the origin $(0, 0)$ and length $m_k |I_{n_0}|$. Attach to its right endpoint a line segment, called a *diagonal*, sloping backwards with angle θ, and length $g(n_1)/\cos\theta$. The diagonal points down if n_1 is even and up if n_1 is odd. Attach to the free end of this diagonal, forming the angle θ a horizontal line segment of length $m_k |I_{n_1}|$. Attach to the free endpoint of this second horizontal line a diagonal of length $g(n_2)/\cos\theta$, again sloping backwards and up or down according to whether n_2 is odd or even. Continue until k horizontal diagonal lines have been drawn, one for each of the intervals I_{n_i}. Finally, draw a diagonal with endpoint the rightmost endpoint of the curve and length $g(0)/\cos\theta$. Call this final curve Q_k. Denote the diagonal with length $g(n)/\cos\theta$ by Δ_n. (See figure 1.)

Notice by the definitions of the constants m_k and $g(0)$, the right endpoint of Q_k has coordinates $(1,0)$: by the construction the left endpoint has coordinates $(0,0)$. Therefore, a closed curve in the cylinder can be made from Q_k via the identification $(x,y) \sim (x+1,y)$. However for simplicity, we work in $\mathbf{R}^2$ and do not close up the endpoints.

Notation. Let $\langle x \rangle$ denote the fractional part of x so that $\langle x \rangle \equiv x$ (mod 1).

Application to continued fractions. For the remainder of the paper, we consider the sequence $x_n = \langle n\alpha \rangle$ where $\alpha \in [0,1)$ is irrational.

Let $a_k \in \mathbf{Z}^+$. Define

$$\frac{p_n}{q_n} = \cfrac{1}{a_1 + \cfrac{1}{a_2 + \cfrac{1}{a_3 + \ldots \cfrac{1}{a_n}}}}$$

where $(p_n, q_n) = 1$. Then $\alpha = \lim_{n\to\infty} p_n/q_n$ exists and is a positive irrational < 1. We write $\alpha = [a_1, a_2, \ldots]$. The fractions p_n/q_n are called *rational convergents* of α. The integers p_n and q_n satisfy the recursive relations:

$$p_{-1} = 1, \quad p_0 = a_0 \quad \text{and} \quad p_n = a_n p_{n-1} + p_{n-2}$$
$$q_{-1} = 0, \quad q_0 = 1 \quad \text{and} \quad q_n = a_n q_{n-1} + q_{n-2}.$$

Define $r_0 = 1$, $r_1 = \alpha$ and

$$r_{n-1} = a_n r_n + r_{n+1}.$$

It is easy to prove that $r_n q_n + r_{n+1} q_{n-1} = 1$.

Let Q denote the fractal plot generated by x_n, θ and g. We call Q a *continued fractal* because Q records the continued fraction expansion of α.

Suppose you have drawn Q_k, for k large. To read off the initial integers in the continued fraction expansion of α, just count the number of the largest monotone decreasing diagonals, sloping both ways, appearing consecutively. This number is a_1. Between two of these segments, count the number of largest monotone decreasing diagonals to get a_2. It is easiest to see with an example:

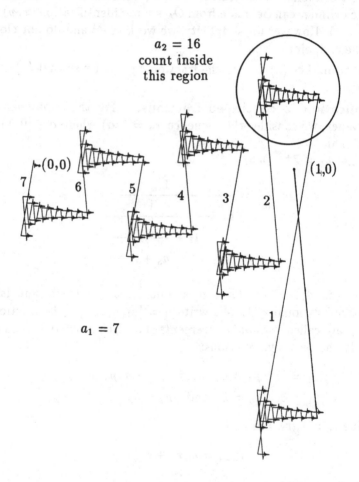

Figure 2 - The continued fractal for $x_n = <n\pi>$, $g(n) = .7/n^{.7}$, $n \le 10,0$
$$a_1 = 7 \quad a_2 = 16$$

Embedded continued fractals. An arc is said to be *embedded* if it has no self-intersections. For $\theta = \pi/2$, the finite fractal plot Q_k is always embedded. If $\theta < \pi/2$, it is difficult for Q_k to be embedded. Self-intersections typically abound. It is quite remarkable that the limit curve Q is *ever* embedded for small θ.

If Q is a continued fractal, number theory influences its structure. For precision, we set $g(n) = 1/n^\gamma$ for $\gamma > 0$. Let $0 < \theta \leq \pi/2$ and $x_n = \langle n\alpha \rangle$. Let Q_k be the finite continued fractal generated by this data. We first prove that if α is poorly approximated by rationals then $Q = \lim Q_k$ exists and is continuous (i.e. connected), for some $\gamma < 1$. The infimum of all such γ which guarantee the continuity of Q depends on how well α is approximated by rationals.

Definition. An irrational number α satisfies a *Diophantine condition* σ if there exist $C > 0$ such that

$$\left| \alpha - \frac{p}{q} \right| > \frac{C}{q^{1+\sigma}} \quad \text{for every rational} \quad \frac{p}{q}.$$

The infimum of all σ such that α satisfies a Diophantine condition σ is the *Diophantine type* of α.

Examples. All algebraic numbers such as the golden mean, $\alpha = (\sqrt{5} + 1)/2$, have Diophantine type 1. See [9], for example.

Theorem 1. *Let $\alpha \in \mathbf{R} \setminus \mathbf{Q}$ have Diophantine type δ and $\epsilon > 0$. Assume $\gamma > 1 - 1/2\delta$. Let Q_k be the sequence of curves determined by $g(n) = 1/n^\gamma$ and α. Then*

$$Q = \lim_{k \to \infty} Q_k$$

exists and is continuous.

For the proof, see Theorem 3.1 of [7].

§2. Quadratic continued fractals.

Since most numbers satisfy a Diophantine condition, Theorem 1 tells us that for almost every α, Q is a connected curve. If α is a quadratic irrational, then we know even more — Q has no self-intersections.

Theorem 2. *Suppose α is a quadratic irrational. There exists $\epsilon > 0$ such that if $1 - \gamma < \epsilon$ then the continued fractal based on $x_n = \langle n\alpha \rangle$ and $g(n) = 1/n^\gamma$ is embedded.*

The proof for $\alpha = \sqrt{2}$ appears as Corollary 3.10 of [6]. It generalises to any quadratic α [7].

It is well known that an irrational number α has a periodic continued fraction iff α is a quadratic irrational. This is reflected in the geometry of the continued fractal of α. This geometry is the subject of the remainder of the paper.

Definitions. For each non-negative real number there is a corresponding s-dimensional Hausdorff measure μ_s, defined as follows: Let $B \subset X$. The zero-dimensional measure $\mu_0 B$ is the number of points in B. For $s > 0$, $\delta > 0$, let

$$\mu_{s,\delta} B = \inf \sum_i [\mathrm{diam}(B_i)]^s$$

where the infimum is taken over all covers $\{B_i\}$ of B such that $\mathrm{diam}(B_i) < \delta$ for each i. Then

$$\mu_s B = \lim_{\delta \to 0^+} \mu_{s,\delta} B.$$

The Hausdorff dimension of a set B is defined by $s = \dim(B) = \inf\{r : \mu_r B = 0\}$. If $\dim(B) = s$ then $\mu_s B$ is the *Hausdorff measure of B within its dimension.*

Suppose $S, T \subset \mathbf{R}^n$. Then T is a *homothetic replica* of S if there exists a mapping $f : \mathbf{R}^n \to \mathbf{R}^n$ of the form $f(x) = ax + b$ where $a \in \mathbf{R}$, $b \in \mathbf{R}^n$ and $f(S) = T$. The real number a is the *expansion constant* of f, b is the *translation constant.*

A curve Q in $\mathbf{R}^2$ is *self-similar* if the following properties hold.
(1) For every $\epsilon > 0$ there exists a covering of Q by closed arcs $Q_{\epsilon,i}$ with diameter satisfying

$$0 < |Q_{\epsilon,i}| \leq \epsilon$$

(2) Each of the arcs $Q_{\epsilon,i}$ is a homothetic replica of Q

This definition is similar to that in Falconer [2] the main difference is that he only requires that the intersection of $Q_{\epsilon,i}$ with $Q_{\epsilon,j}$ to have s-measure 0.

A curve Q in $\mathbf{R}^2$ is *asymptotically self-similar* if there exists $Q' \subset \mathbf{R}^2$ such that

(1) For every $\epsilon > 0$ there exists a covering of Q by closed arcs $Q_{\epsilon,i}$ with diameter

$$0 < |Q_{\epsilon,i}| \leq \epsilon;$$

(2) For each of the arcs $Q_{\epsilon,i}$ there exists a homothetic replica $Q'_{\epsilon,i}$ and $Q'_{\epsilon,i} \to Q'$ for $\epsilon \to 0$.(Convergence is measured with the Hausdorff metric.)

An embedded arc Q contained in a metric space (X, d) is called a *quasi-arc* if there exists a constant $K > 0$ such that if x, $y \in Q$ then the subarc of Q connecting x and y is contained in a disc of diameter $K d(x, y)$.

Theorem 3. *Let α be a quadratic irrational. There exist constants $C_0 > 0$ and $0 < \gamma_0 < 1$ such that if $0 < C < C_0$ and $\gamma_0 < \gamma$ then the fractal plot Q built from $x_n = \langle n\alpha \rangle$ and $g(n) = C/n^\gamma$ satisfies*

(1) Q is a quasi-arc;
(2) The Hausdorff dimension of Q is $1/\gamma$. The Hausdorff measure of Q within its dimension is positive;
(3) Q is asymptotically self-similar.

The proofs to (1) and (2) appear in [7]. Part (3) is proved in the appendix.

Computer experiments on non-quadratic algebraic numbers. It is not known if there is any pattern to the continued fraction expansion of algebraic numbers which are not quadratic. This is one of the oldest problems in mathematics. The continued fractal contains all the information about α and so might indicate geometrically any pattern in its continued fraction expansion.

In the next figures, note there is some sort of self-similarity, at least for $n < 10^5$. It remains to be seen if these patterns will repeat with even more iterates and blow-ups. It should be noted that the plots for simple quadratic numbers such as $\sqrt{2}$, $\sqrt{3}$ show their self-similarity quickly — you do not have to look at very fine detail plots for simple quadratic numbers such as $\sqrt{2}$, $\sqrt{3}$ show their self-similarity quickly — you do not have to look at very fine detail to see the self-similarity. The hope is that a geometric pattern might show up quickly for the simplest third or fourth roots, for example. Otherwise, the experiment might not be viable.

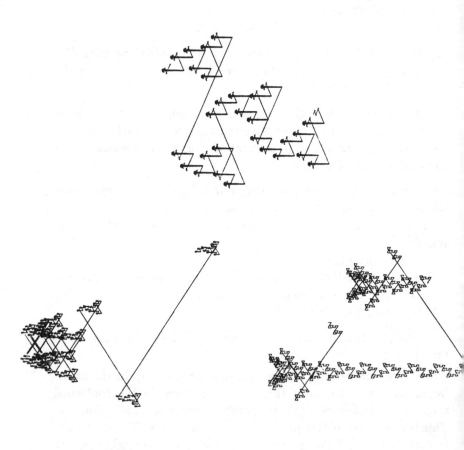

Figure 3 — The continued fractal for $x_n = \,<n\alpha>$, $\alpha = 2^{\frac{1}{4}}$, $g(n) = .7/n$
$n \leq 10,000$ and details from $1/100$ and $1/10000$ the original

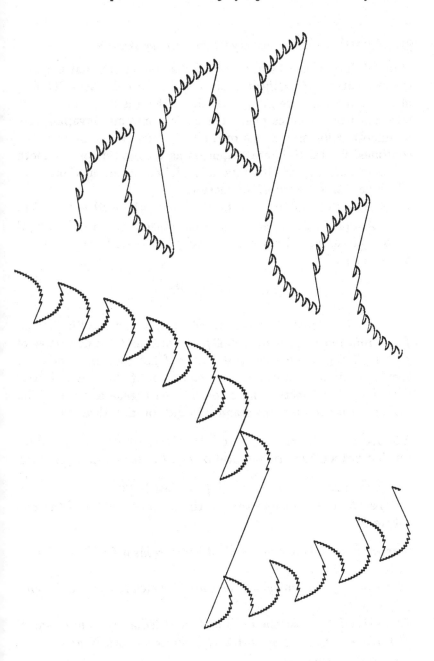

Figure 4 — The continued fractal for $x_n = <n\alpha>$, $\alpha = 7^{\frac{1}{4}}$
$g(n) = .2/n^{.8}$, $n \leq 8,000$ and detail from 1/100 the original

§3. Applications to dynamical systems.

Fractal Jordan curves appear as Julia sets of rational maps of the two-sphere S^2. Diffeomorphisms of S^2 are dynamically simpler because they are invertible. Since there is no "folding" our intuition from calculus might tell us that minimal invariant sets of smooth diffeomorphisms ought to be smooth. However, the continued fractal of figure 1 appears as a Julia set for a smooth diffeomorphism of S^2. It contains a minimal invariant Cantor set which has large Hausdorff dimension.

Define the annulus A to be $S^1 \times [0,1]$ where $S^1 \simeq \mathbf{R}/\mathbf{Z}$. Let $M \subset A$ and $f : M \to M$ a homeomorphism. Let $\tilde{A} \simeq \mathbf{R} \times [0,1]$ denote the universal cover of A and $\tilde{f} : \tilde{M} \to \tilde{M}$ the lift of $f :$ $M \to M$. Define

$$\rho(f, x) = \lim_{n \to \infty} \frac{\tilde{f}_1^n(x)}{n}$$

If $\alpha = \rho(f, x)$ exists and is indepedent of x and the lift $\tilde{f}$, then f has *rotation number* α. Let Df_x denote the total derivative of f at x. If $Df_x =$ Identity, $x \in M$ and f has rotation number α, then f is said to be a *generalised rotation* of M through α. Notice that if M is connected and smooth, then a generalised rotation through α is precisely the standard rigid rotation through α.

Theorem 4. *There exists a C^2 irrational rotation of a Cantor set, but not a C^3 irrational rotation of a Cantor set in a quasi-arc.*

The first part is proved in [6], the last in [3].

Let N and S denote the North and South poles of the two sphere S^2.

Theorem 5. *There exists a C^2 diffeomorphism $f : S^2 \to S^2$ such that*
(1) The only periodic points of f are N which repels and S which attracts;
(2) There is at least one orbit of f bounded away from N and S
(3) There is at least one orbit asymptotic to both N and S.

This is proved in [5]. The orbit bounded away from N and S is a Cantor set contained in the continued fractal illustrated in Figure 1.

Theorem 6. *If there exists a C^r diffeomorphism of S^2 satisfying (1)-(3) above,then there exists a C^r vector field on S^3 with no zeros and no closed integral curves.*

The proof of this theorem is the subject of [8] (see also [5]). Theorems 5 and 6 lead to a C^2 counter example to the Seifert Conjecture:

Corollary. *There exists a C^2 vector field on S^3 with no zeros and no closed integral curves.*

For an overview of the proof see [4].

§4. Appendix

A continued fractal Q contains a canonical Cantor set Γ, the closure of the set of endpoints of the diagonal segments of Q.

Theorem. *The Cantor set Γ of the continued fractal Q generated by $x_n = \langle n\alpha \rangle$ and $g(n) = 1/n^\gamma$ is asymptotically self-similar if α is a quadratic irrational.*

Proof. Recall the notation for continued fractions. Since α is a quadratic irrational, there exists a minimal integer p, called the period of α, such that $\alpha = [a_1, a_2, \ldots, a_p, a_1, a_2, \ldots, a_p, a_1, a_2, \ldots, a_p, \ldots]$. That is, $a_{k+i} = a_i$ if k is a multiple of p.
Hence,

$$q_{k+i} = a_i q_{k-1+i} + q_{k-2+i}, \qquad 0 \le i \le p.$$

Define $G : I \to I$ by $G(x) = 1/x \pmod 1$. Then $G^p(\alpha) = \alpha$. Let $\alpha_n = G^n(\alpha)$. Note that $\alpha_{k+i} = \alpha_i$ if k is a multiple of p.
Let W_k denote the set of intervals of I complementary to

$$\{\langle 0\alpha \rangle, \langle 1\alpha \rangle, \langle 2\alpha \rangle, \ldots, \langle (q_k + q_{k-1} - 1)\alpha \rangle\}.$$

Assume k is a multiple of p. There is a mapping $h : I \to \mathbf{R}^2$ such that $h(I) = Q$. For U an open interval in W_k, let Q_U denote the closure of $h(U)$. Notice that Q_U does not contain the diagonals at either end.
Next we estimate the diameter of Q_U. We may assume $|U| = r_k$, otherwise U is an element of W_{k+1}. (A proof of this standard fact is contained in [6]. It follows from the relation

$q_k r_k + q_{k-1} r_{k+1} = 1$). Let m be the minimal index of points $\langle n\alpha \rangle$ appearing in U. It follows that $q_k < m$. Let π_i, $i = 1, 2$ denote the projections onto the first and second coordinates in $\mathbf{R}^2$. Then

$$|\pi_1 Q_U| = \lim_{n \to \infty} \left| m_n |U| - \sum_{i=q_k+1}^{q_n} \chi_U \langle i\alpha \rangle g(i) \right|$$

$$= \left| |U| \left(1 + \sum_{i=0}^{q_k} g(i) \right) - \sum_{i=q_k+1}^{\infty} (\chi_U \langle i\alpha \rangle - |U|) g(i) \right|$$

$$< \frac{r_k C q_k^{1-\gamma}}{1 - \gamma} + \frac{2C}{q_k^\gamma} \qquad \text{by Theorem 2.4(i) of [6]}$$

$$< \frac{C}{q_k^\gamma (1 - \gamma)} + \frac{2C}{q_k^\gamma} \qquad \text{since } r_k q_k < 1$$

$$= \left(\frac{C}{1 - \gamma} + 2C \right) \frac{1}{q_k^\gamma}.$$

The second coordinate is similar:

$$|\pi_2 Q_U| = \lim_{n \to \infty} \left| \sum_{q_k < 2i+1 \leq q_{k+n}} \chi_U \langle (2i+1)\alpha \rangle g(2i+1) \right.$$

$$\left. - \sum_{q_k < 2i \leq q_{k+n}} \chi_U \langle 2i\alpha \rangle g(2i) \right|$$

$$\leq \left| \sum_{q_k < 2i+1} (\chi_U \langle (2i+1)\alpha \rangle - |U|) g(2i+1) \right|$$

$$- \left| \sum_{q_k < 2i} (\chi_U \langle 2i\alpha \rangle - |U|) g(2i) \right|$$

$$< \frac{4C}{q_k^\gamma} \qquad \text{by Theorem 2.4(i) of [6]}.$$

For $\epsilon > 0$, fix k so that

$$\left(6C + \frac{C}{1 - \gamma} \right) \frac{1}{q_k^\gamma} < \epsilon.$$

It follows immediately that $\{Q_U : U \in W_k\}$ covers Γ and the diameter of the cover is $\leq \epsilon$.

Let Q'_U be the homothetic replica of Q_U with expansion constant $a = q_k^\gamma$ and translation constant b so that the left endpoint is the origin. We consider sequences of the $\{Q'_U\}$.

We need three lemmas:

Lemma 1. *If $\alpha \in \mathbf{R} \setminus \mathbf{Q}$ then $r_n = \alpha_0 \alpha_1 \ldots \alpha_{n-1}$.*

Proof. By definition, $r_0 = 1, r_1 = \alpha = \alpha_0$. Assume the statement for r_k and r_{k-1}. Then

$$
\begin{aligned}
r_{k+1} &= r_{k-1} - a_k r_k \\
&= \alpha_0 \alpha_1 \ldots \alpha_{k-2} - a_k \alpha_0 \alpha_1 \ldots \alpha_{k-1} \\
&= \alpha_0 \alpha_1 \ldots \alpha_{k-2}(1 - a_k \alpha_{k-1}) \\
&= \alpha_0 \alpha_1 \ldots \alpha_{k-2} \alpha_{k-1}(1/\alpha_{k-1} - a_k) \\
&= \alpha_0 \alpha_1 \ldots \alpha_k.
\end{aligned}
$$

Lemma 2. *If $\alpha \in \mathbf{R} \setminus \mathbf{Q}$ then*

$$
\begin{aligned}
r_{n+1} q_{n+1} = {} & \alpha_n \alpha_{n-1} - \alpha_n \alpha_{n-1}^2 \alpha_{n-2} \\
& + \alpha_n \alpha_{n-1}^2 \alpha_{n-2}^2 \alpha_{n-3} \\
& - \alpha_n \alpha_{n-1}^2 \alpha_{n-2}^2 \alpha_{n-3}^2 \alpha_{n-4} + \cdots \\
& (-1)^{n+1} \alpha_n \alpha_{n-1}^2 \alpha_{n-2}^2 \alpha_{n-3}^2 \cdots \alpha_0.
\end{aligned}
$$

Proof. By Lemma 1, $r_{n+1}/r_{n-1} = \alpha_{n-1} \alpha_n$. Therefore

$$
\begin{aligned}
r_n q_n &= 1 - r_{n-1} q_{n+1} \\
&= 1 - \frac{r_{n+1} q_{n+1}}{\alpha_{n-1} \alpha_n}.
\end{aligned}
$$

Thus

$$
\begin{aligned}
r_{n+1} q_{n+1} &= \alpha_n \alpha_{n-1}(1 - r_n q_n) \\
&= \alpha_n \alpha_{n-1}(1 - \alpha_{n-1} \alpha_{n-2}(1 - r_{n-1} q_{n-1})) \\
&= \alpha_n \alpha_{n-1} - \alpha_n \alpha_{n-1}^2 \alpha_{n-2}(1 - r_{n-1} q_{n-1}) \\
&= \alpha_n \alpha_{n-1} - \alpha_n \alpha_{n-1}^2 \alpha_{n-2} + \cdots \\
& (-1)^{k+2} \alpha_n \alpha_{n-1}^2 \alpha_{n-2}^2 \cdots \alpha_{n-k}^2 \alpha_{n-k-1}(1 - r_{n-k} q_{n-k}).
\end{aligned}
$$

If $n = k$ then $r_{n-k} q_{n-k} = 1$. The result follows.

Lemma 3. *If α is a quadratic irrational with period p then $\{r_k q_k :$ k is a multiple of $p\}$ is a Cauchy sequence.*

Proof. By Lemma 2, if $m > n$ and both are multiples of p,

$$|r_n q_n - r_m q_m| = |\alpha_{m-1}\alpha_{m-2}^2 \ldots \alpha_{m-n}^2 \alpha_{m-n-1} + \ldots$$
$$(-1)^m \alpha_{m-1}\alpha_{m-2}^2 \ldots \alpha_1^2 \alpha_0|$$
$$= AB$$

where

$$A = \alpha_{m-1}\alpha_{m-2}^2 \ldots \alpha_{m-n}^2 \alpha_{m-n-1} = \alpha_{n+p}\alpha_{n+p-1}^2 \ldots \alpha_p^2 \alpha_{p-1}$$

and

$$B = 1 - \alpha_{m-n-1}\alpha_{m-n-2} + \alpha_{m-n-1}\alpha_{m-n-2}^2 \alpha_{m-n-3}$$
$$- \alpha_{m-n-1} \ldots \pm \alpha_1^2 \alpha_0.$$

Since the α_i are bounded away from 0 and 1, it follows that $A \to 0$ as $n \to \infty$. Since $0 < B < 1$, we conclude that $r_k q_k$ is Cauchy.

We are now ready to show that any sequence of Q'_U converges.

First parametrise Q'_U. Define a mapping $f_U : I \to Q'_U$ by $f_U(x) = a(h(xr_k + p)) + b$ where p is the left endpoint of U.

We estimate $|\pi_i f_V(x) - \pi_i f_U(x)|$, $i = 1, 2$, where $U \in W_k$ and $V \in W_j$.

Let $I_U = [p, p + xr_k)$ for $x \in I$. Then

$$\pi_1 f_U(x) = a \lim_{n \to \infty} \left[m_n x r_k - \sum_{i=q_k+1}^{q_n} \chi_{I_U} \langle i\alpha \rangle g(i) \right]$$

$$= \lim_{n \to \infty} q_k^\gamma \left[x r_k \left(1 + \sum_{i=0}^{q_k} g(i) \right) \right.$$

$$\left. - \sum_{i=q_k+1}^{q_n} (\chi_{I_U} \langle i\alpha \rangle - |I_U|) g(i) \right].$$

Now

$$|\pi_1 f_V(x) - \pi_1 f_U(x)| \le \lim_{n \to \infty} x \left| r_k q_k^\gamma \sum_{i=0}^{q_k} \frac{1}{i^\gamma} - r_j q_j^\gamma \sum_{i=0}^{q_j} \frac{1}{i^\gamma} \right|$$

$$+ \left| \sum_{i=q_j+1}^{q_n+j} (\chi_{I_V} \langle i\alpha \rangle - |I_V|) \frac{q_j^\gamma}{i^\gamma} \right.$$

$$\left. - \sum_{i=q_k+1}^{q_n+k} (\chi_{I_U} \langle i\alpha \rangle - |I_U|) \frac{q_k^\gamma}{i^\gamma} \right|.$$

The difference of the first two series is $< 2x(r_k q_k - r_j q_j)/(1 - \gamma)$. By Lemma 3, this tends to 0 as $j, k \to \infty$.

We estimate the difference of the last two series. Observe that

$$\sum_{i=q_k+1}^{q_n+k} \chi_{I_U} \langle i\alpha \rangle = \sum_{i=q_j+1}^{q_n+j} \chi_{I_V} \langle i\alpha \rangle.$$

This follows since α is quadratic and k and j are both multiples of p.

Renormalisation. The pattern of points $< i\alpha >$ appearing in I_U is exactly the same as that in I_V within the given bounds. Note that $i = \sum b_m q_m$ where $0 \le b_m \le a_m$. Also, if $< i\alpha > \in V$ iff $< i'\alpha > \in U$ where $i' = \sum b_m q_{m+j+k}$.

Moreover, if $j \ge k$, there exists a unique I_W such that

(1) $\frac{|I_V|}{|I_W|} = |I_U|$

(2) $I_V \subset I_W$

(3) the points in I form a homothetic replica of the points in I_W with $\langle i\alpha \rangle \in I_W$ corresponding to $\langle i'\alpha \rangle \in I$.

Hence

$$\sum_{i=q_k+1}^{q_{n+k}} (\chi_{I_U}\langle i\alpha\rangle - |I_U|)\frac{1}{i^\gamma} = \sum_{i=q_j+1}^{q_{n+j}} (\chi_{I_W}\langle i\alpha\rangle \left(\chi_{I_V}\langle i\alpha\rangle - \frac{|I_V|}{|I_W|}\right)\frac{1}{i'^\gamma}$$

$$= \sum_{i=q_j+1}^{q_{n+j}} \left(\chi_{I_V}\langle i\alpha\rangle - \frac{\chi_{I_W}|I_V|}{|I_W|}\right)\frac{1}{i'^\gamma}$$

$$= \sum_{i=q_j+1}^{q_{n+j}} (\chi_{I_V}\langle i\alpha\rangle - |I_V|)\frac{1}{i'^\gamma}$$

$$+ \frac{|I_V|}{|I_W|}\sum_{i=q_j+1}^{q_{n+j}} (|I_W| - \chi_{I_W}\langle i\alpha\rangle)\frac{1}{i'^\gamma}.$$

The absolute value of the last series is here bounded by $2|I_V|/|I_W| = 2|I_U|$ by 2.4(1) of [6].

It follows that the difference of the second series of $|\pi_1 f_V(x) - \pi_1 f_U(x)|$ is bounded by

$$\left|\sum_{i=q_j+1}^{\infty} (\chi_{I_V}\langle i\alpha\rangle - |I_V|)\left(\frac{q_j^\gamma}{i^\gamma} - \frac{q_k^\gamma}{i'^\gamma}\right) + 2|I_U|\right| <$$

$$2\left|\frac{q_j^\gamma}{(q_{j+1})^\gamma} - \frac{q_k^\gamma}{(q_{k+1})^\gamma}\right| + 2r_k.$$

The last sum clearly $\to 0$ as $j, k \to \infty$.

Last of all, we estimate $|\pi_2 f_U(x) - \pi_2 f_V(x)|$:

$$\pi_2 f_U(x) = \sum_{q_k < 2i+1 \leq q_{n+k}} \chi_{I_U} \langle (2i+1)\alpha \rangle \frac{q_k^\gamma}{(2i+1)^\gamma}$$

$$- \sum_{q_k < 2i \leq q_{n+k}} \chi_{I_U} \langle 2i\alpha \rangle \frac{q_k^\gamma}{(2i)^\gamma}$$

$$= \sum_{q_k < 2i+1 \leq q_{n+k}} (\chi_{I_U} \langle (2i+1)\alpha \rangle - |I_U|) \frac{q_k^\gamma}{(2i+1)^\gamma} -$$

$$\sum_{q_k < 2i \leq q_{n+k}} (\chi_{I_U} \langle 2i\alpha \rangle - |I_U|) \frac{q_k^\gamma}{(2i)^\gamma}$$

$$+ q_k^\gamma |I_U| \sum_{q_k < 2i \leq q_{n+k}} \frac{1}{(2i)^\gamma} - \frac{1}{(2i+1)^\gamma}.$$

Now $|\pi_2 f_U(x) - \pi_2 f_V(x)|$ consists of these three sums together with three corresponding sums for f_V. The four series involving the characteristic function χ are estimated in the same way as for $|\pi_1 f_U(x) - \pi_1 f_V(x)|$. They tend to 0 as $k, j \to \infty$. The last two are bounded by $\frac{2r_k}{q_k} + \frac{2r_j}{q_j} \to 0$ as $j, k \to \infty$. Hence $\{Q'_U\}_k$ is a Cauchy sequence.

§5. References

[1] F. M. Dekking and M. Mendes-France, 'Uniform distribution modulo one; geometric point of view' *J. Reine Angew. Math.*, **329** (1981) 143–153.

[2] K. J. Falconer, *The Geometry of Fractal Sets*, Cambridge University Press (1985).

[3] J. Harrison, 'Dynamics on Ahlfors quasi-circles', *Proc. Ind. Phil. Soc.*, to appear (1989).

[4] J. Harrison, 'Continued fractals and the Seifert conjecture',*Bulletin A.M.S.*, **13** (1985) 147–153.

[5] 'C^2 Counterexamples to the Seifert conjecture', *Topology*, **27** (1988) 49-78.

[6] J. Harrison, 'Denjoy fractals', *Topology*, to appear (1989).

[7] J. Harrison, 'Embeddings of continued fractals and their Hausdorff dimension', *Constructive Approximation*, to appear (1989).

[8] J. Harrison, 'The loxodromic mapping problem', *Journal of Differential Equations*, to appear (1989).

[9] L. Kuipers and H. Niederreiter *Uniform distribution of sequences* (Wiley, New York 1974).

9
Chaos implies confusion

Michel Mendes France

Université de Bordeaux 1, Talence, France

§1. A dynamical system and transcendental numbers

Consider the baker's transformation T which maps the unit square onto itself. Flatten the unit square to obtain the $2 \times 1/2$ rectangle then fold the rectangle onto itself to reproduce the square.

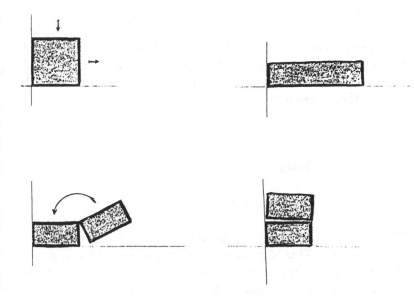

Figure 1

(This transformation is actually closer to the real baker transformation than that which ergodicians usually discuss). We now repeat again and again the transformation T thus defining the iterates $(T^2, T^3, \ldots)$.

We study this well known transformation in a somewhat unconventional way. We shall only be concerned by the "squeleton" of T. Imagine a sheet of paper folded upon itself n times. At each folding operation,

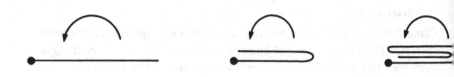

Figure 2

we suppose that the sheet of paper is stretched to twice its length. Unfolding the sheet we read off the pattern formed by $2^n - 1$ folds which are of two kinds, either "valleys" ∨ or "mountains" ∧. For $n = 3$, the sequence is

$$\vee \vee \wedge \vee \vee \wedge \wedge.$$

For $n = 4$, we obtain

$$\vee \vee \wedge \vee \vee \wedge \wedge \vee \vee \vee \wedge \wedge \vee \wedge \wedge.$$

The first $2^3 - 1 = 7$ terms of the above sequence coincide with the first sequence. This observation is of course obvious and general: the $2^n - 1$ first terms of the sequence obtained after $n + 1$ folds is the sequence obtained after n folds. It thus makes sense to let n go to infinity and to consider the infinite paperfolding sequence

$$\vee \vee \wedge \vee \vee \wedge \wedge \vee \vee \vee \wedge \wedge \vee \wedge \wedge \vee \vee \ldots$$

This sequence contains all the information of the baker's transformation.

Instead of folding the sheet of paper (or the dough) in the positive direction, we may decide at each step to fold at will either in the positive direction or in the negative direction (see fig. 3).

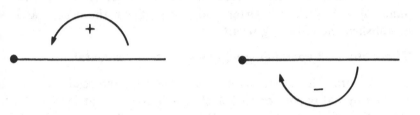

Figure 3

Let X be the space of finite or infinite sequences of $\wedge$'s and $\vee$'s:

$$X = \left[\bigcup_{n=1}^{\infty} \{\wedge, \vee\}^n \right] \cup \{\wedge, \vee\}^{\mathbf{N}},$$

together with the metric d defined as follows. Let $A \in X$, $B \in X$,

$$A = \{V_1, V_2, V_3, \ldots\} \qquad B = \{V_1', V_2', V_3' \ldots\}.$$

Then

$$d(A, B) = (\inf\{n \geq 1 : V_n \neq V_n'\})^{-1}$$

and X is a metric complete space. Thus, folding a sheet of paper infinitely many times with arbitrary folding instructions at each step will always generate at least one infinite sequence which we shall again call a paperfolding sequence. The family of paperfolding sequences has the power of the continuum.

Let $\{V_1, V_2, V_3, \ldots\}$ be a paperfolding sequence. Then the subsequence $\{V_{2n-1}\}$ is the alternating sequence $\vee \wedge \vee \wedge \ldots$ or $\wedge \vee \wedge \vee \ldots$ and $\{V_{2n}\}$ is a paperfolding sequence. This property is actually a characterisation of paperfolding sequences.

Suppose we map $\vee$ onto 1 and $\wedge$ onto 0. A paperfolding sequence is now an infinite sequence of 0's and 1's which can be considered as the binary digit expansion of a real number

$$0.V_1V_2V_3\ldots = \sum_{n=1}^{\infty} \frac{V_n}{2^n},$$

a paperfolding number. Loxton and van der Poorten on the one hand [5] and van der Poorten and myself [9] on the other hand established the following result.

Theorem 1. *Paperfolding numbers are transcendental.*

The proof is rather involved and we refer the reader either to the original papers or to Folds! which discusses at length the properties of these sequences. Before closing this section let us mention a simple but surprising fact discovered by D. Knuth and C. Davis [1].

Theorem 2. *Put* $\vee = 1$ *and* $\wedge = -1$. *Paperfolding sequences which begin by* $\vee$ *are multiplicative, i.e. for coprime integers m,n,*

$$V_{mn} = V_m V_n$$

§2. Dragon curves

The infinitely folded sheet of paper is now unfolded to 90°. The edge of the sheet of paper is a polygonal line drawn on a square lattice, Z^2 say. The basic property of these polygons called dragon curves [1] was first described by C. Davis and D. Knuth.

Theorem 3. *Dragon curves are self avoiding.*

In this context, self avoiding means that the curves are simple: each edge of the Z^2–lattice is visited at most once by a given dragon curve. Vertices however may be visited twice at most.

Figure 4 represents the beginning the dragon curve obtained by positive foldings. Figure 5 corresponds to the instructions $+ - + - + \ldots$

Dragon curves seem quite erratic and suggest lattice filling. We shall see in the following paragraphs that indeed, their "entropy" is relatively high and that their "dimension" is 2.

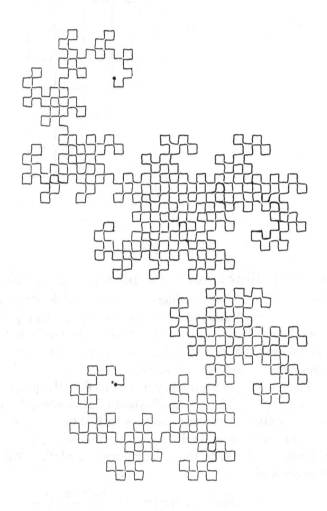

Figure 4

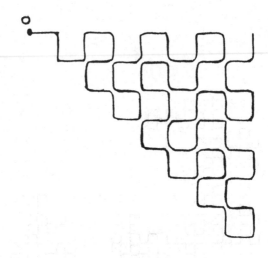

Figure 5

§3. The dimension of a planar curve [2],[10]

All the curves Γ we consider are locally rectifiable, unlike Weierstrass curves, Peano curves or fractal curves. Our point of view, even though influenced by the work of Mandelbrot [6], is quite independent. We assume that the curves Γ are infinite and that every compact set of the plane contains at most a finite portion of Γ. The curves are necessarily unbounded in the plane.

Let $\epsilon > 0$. The ϵ magnification of Γ, denoted by $\Gamma(\epsilon)$, is the set of points of which the distance to Γ is less than ϵ.

For positive s, let Γ_s represent the beginning portion of Γ of length s; $\Gamma_s(\epsilon)$ is the ϵ–magnification of Γ_s. We define the dimension of Γ:

$$\dim{}^* \Gamma = \lim_{\epsilon \to 0} \limsup_{s \to \infty} \frac{\log \operatorname{area} \Gamma_s(\epsilon)}{\log \operatorname{diam} \Gamma_s}$$

$$\dim_* \Gamma = \lim_{\epsilon \to 0} \liminf_{s \to \infty} \frac{\log \operatorname{area} \Gamma_s(\epsilon)}{\log \operatorname{diam} \Gamma_s}$$

where $\operatorname{diam}(\Gamma_s)$ is the diameter of the point set Γ_s. It is easy to see that

$$1 \leq \dim_* \Gamma \leq \dim{}^* \Gamma \leq 2$$

and that for all $\alpha, \beta \in \mathbf{R}$ such that $1 \leq \alpha \leq \beta \leq 2$ there exists a curve Γ with

$$\dim_* \Gamma = \alpha \qquad \dim^* \Gamma = \beta.$$

If $\dim_* \Gamma = \dim^* \Gamma$, we note the common value by $\dim \Gamma$.

A straight line has dimension 1. Curves which go "rapidly" to infinity such as the spiral $\rho = \exp \theta$ are also one-dimensional. The spiral $\rho = \theta^\alpha$ has dimension

$$\min\{2, 1 + 1/\alpha\} \qquad (\alpha > 0).$$

In particular, when $\alpha \leq 1$, the spiral is bidimensional.

§4. Resolvable curves

The dimension of a given curve Γ may be difficult to compute if only because of the difficulty in measuring the area of $\Gamma_s(\epsilon)$. If however, the curve Γ does not "bunch up" too much, the area is approximately $2\epsilon s$. A curve is said to be *resolvable* if there exists an $\epsilon_0 > 0$ for which

$$\liminf_{s \to \infty} \frac{\text{area}\,\Gamma_s(\epsilon_0))}{s} > 0$$

Then the condition holds for all $\epsilon \in \,]0, \epsilon_0]$. The spiral $\rho = \theta^\alpha$ is resolvable if and only if $\alpha \geq 1$.

For a resolvable curve,

$$\text{area}\,\Gamma_s(\epsilon) > a(\epsilon)s$$

for all large s and all small $\epsilon > 0$; here $a(\epsilon)$ is some strictly positive constant. Calculating the dimension is then simplified since

$$\dim^* \Gamma = \limsup_{s \to \infty} \frac{\log s}{\log \operatorname{diam} \Gamma_s}$$

$$\dim_* \Gamma = \liminf_{s \to \infty} \frac{\log s}{\log \operatorname{diam} \Gamma_s}.$$

Dragon curves are resolvable. The above formulas enable one to compute their dimension.

Theorem 4. *All dragon curves are 2 dimensional.*

For the proof we refer to an article of Tenenbaum and myself [10]. The proof shows a close link between paperfolding and the so called Rudin-Shapiro sequences.

§5. Geometric probability

In this paragraph, Γ will always denote a finite curve in the plane. Its length is represented by $|\Gamma|$. Let K and ∂K respectively be the convex hull of Γ and its boundary.

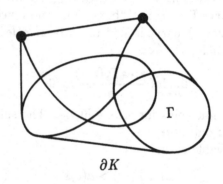

Figure 6

Lengths $|\Gamma|$ and $|\partial K|$ are related by the inequality

$$2|\Gamma| \geq |\partial K|.$$

Equality holds if and only if Γ is a straight segment.

An infinite straight line L is defined by the two polar coordinates ρ,θ where ρ is the distance of the origin to L and where θ is the polar angle.

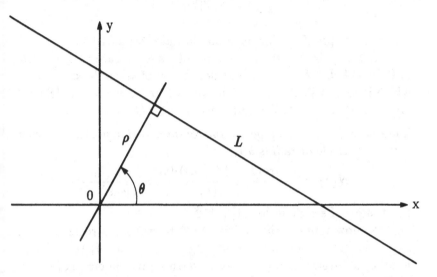

Figure 7

Identification of (ρ,θ) with $(-\rho,\theta+\pi)$ gives a Möbius structure to the set M of infinite straight lines L.

Let $\Omega \subset M$ be the subset of lines L which intersect Γ. We consider Ω as a probability space with measure p defined by

$$dp = \frac{d\rho\, d\theta}{\mathrm{meas}(\Omega)}.$$

For all integers $n \geq 1$, put

$$p_n = \Pr\{\mathrm{card}(L \cap \Gamma) = n\}$$

i.e., p_n is the probability for a random straight line L to intersect Γ in exactly n points. Hence

$$\sum_{n=1}^{\infty} p_n = 1.$$

We shall use the well known result of Steinhaus [12]

Theorem 5. *The average number of intersection points of* Γ *with a random straight line L is:*

$$\sum_{n=1}^{\infty} n p_n = 2|\Gamma|/|\partial K|$$

For the proof see for example Santaló [11] p.31.

At this point we should also recall a special case of a theorem of Poincaré. Let $\delta > 0$ and consider the set of all circles of radius δ which intersect Γ. A circle with radius δ and center m is denoted $\Delta(\delta, m)$. It intersects Γ if and only if $m \in \Gamma(\delta)$.

Theorem 6. *The average number of intersection points of* Γ *with a random circle of radius δ is*

$$N(\delta) = \frac{\int \operatorname{card}(\Gamma \cap \Delta(\delta, m))\, dm}{\operatorname{meas} \Gamma(\delta)} = \frac{4\delta|\Gamma|}{\operatorname{meas} \Gamma(\delta)}$$

For the proof see Santaló [11] p.112.

We now suppose that Γ is an infinite curve and recall that Γ_s is the beginning portion of Γ of length s. Call $N_s(\delta)$ the average number of intersection points of Γ_s with a random circle of radius δ. Then

$$\limsup_{s \to \infty} N_s(\delta) < \infty$$

if and only if Γ is resolvable.

Notice also that Poincaré's formula shows that

$$\operatorname{meas} \Gamma_s(\delta) = \frac{4\delta s}{N_s(\delta)}$$

so that

$$\dim{}^{*}\Gamma = \lim_{\delta \to \infty} \limsup_{s \to \infty} \frac{\log 4\delta s - \log N_s(\delta)}{\log \operatorname{diam} \Gamma_s}$$

$$\leq \limsup_{s \to \infty} \frac{\log s}{\log \operatorname{diam} \Gamma_s} - \lim_{\delta \to 0} \liminf_{s \to \infty} \frac{\log N_s(\delta)}{\log \operatorname{diam} \Gamma_s},$$

and similarly

$$\dim{}_{*}\Gamma \geq \liminf_{s \to \infty} \frac{\log s}{\log \operatorname{diam} \Gamma_s} - \lim_{\delta \to 0} \limsup_{s \to \infty} \frac{\log N_s(\delta)}{\log \operatorname{diam} \Gamma_s}$$

The speed with which $N_s(\delta)$ increases as s goes to infinity is thus linked to the dimension of Γ. This relationship will be made clearer in paragraph 9.

§6. Entropy of a finite curve [4], [7], [8].

Let us consider once again a finite length curve Γ. By definition its entropy is

$$S(\Gamma) = \sum_{n=1}^{\infty} p_n \log \frac{1}{p_n}$$

where as usual $0.\infty = 0$. Notice that $S(\Gamma) \geq 0$ and that $S(\Gamma) = 0$ if and only if Γ is either a straight segment or a closed convex curve. The entropy measures the complexity of Γ. Complicated curves have high entropy.

Theorem 7. *The entropy of Γ verifies the inequality*

$$S(\Gamma) \leq S_t(\Gamma) := \log \frac{2|\Gamma|}{|\partial K|} + \frac{\beta}{e^\beta - 1}$$

where

$$\beta = \log \frac{2|\Gamma|}{2|\Gamma| - |\partial K|} > 0.$$

Proof: Maximize the function

$$S(p_1, p_2, \ldots) = \sum_{n=1}^{\infty} p_n \log \frac{1}{p_n}$$

where the variables $p_1, p_2, \ldots$ are subject to the two conditions

$$\sum_{n=1}^{\infty} p_n = 1$$

$$\sum_{n=1}^{\infty} n p_n = \frac{2|\Gamma|}{|\partial K|}.$$

The Lagrange technique shows that the maximum is attained for the Gibbs distribution

$$p_n = \hat{p}_n = a \exp(-\beta n),$$

where a and β are two constants related by the equations

$$a = \exp \beta - 1$$

and
$$\beta = \log \frac{2|\Gamma|}{2|\Gamma| - |\delta K|}.$$

Hence
$$\hat{p}_n = \frac{|\partial K|}{2|\Gamma| - |\partial K|} \left(1 - \frac{|\partial K|}{2|\Gamma|}\right)^n.$$

Finally
$$S(\hat{p}_1, \hat{p}_2, \ldots) = S_t(\Gamma). \qquad \text{Q.E.D.}$$

The quantity $S_t(\Gamma)$ should be thought of as the topological entropy of Γ. It is much coarser than the "metric" entropy $S(\Gamma)$ since it depends only on the ratio $|\Gamma|/|\partial K|$.

§7. Thermodynamics [4], [7].

Physicists will recognize the exponent β in the formula $\hat{p}_n = a \exp(-\beta n)$ as being the inverse of the absolute temperature T. We thus identify β^{-1} with the "temperature of the curve Γ":

$$T = \left(\log \frac{2|\Gamma|}{2|\Gamma| - |\partial K|}\right)^{-1}$$

Notice that $T > 0$ as it should!

We pursue the identification by defining the volume V of Γ as its length: $V = |\Gamma|$. The ratio $1/|\partial K|$ is to be identified with the pressure P: the higher the pressure, the smaller the hull of Γ.

Temperature, volume and pressure are thus related by the formula

$$T = \left(\log \frac{2V}{2V - P^{-1}}\right)^{-1}$$

or

$$PV = \frac{1}{2}\left(1 - \exp\left(-\frac{1}{T}\right)\right)^{-1}$$

When T decreases to 0, the product PV tends to $1/2$, i.e. $|\Gamma|/|\partial K| = 1/2$. At 0 degrees, curves freeze to straight segments, the entropy of which are 0. Nernst's law thus applies to curves.

When T increases to infinity,

$$PV \sim \frac{1}{2}T.$$

We recognize Boyle's law which proves that at high temperature, curves behave like a perfect gas.

§8. Entropy of unbounded curves

Let us come back to the study of the entropy of curves. Suppose Γ is an infinite curve and consider the beginning portion Γ_s. In paragraph 6 we defined its entropy $S(\Gamma_s)$ and the "topological" entropy $S_t(\Gamma_s)$. We normalize both entropies by dividing by $\log 2s$:

$$\frac{S(\Gamma_s)}{\log 2s} \le \frac{S_t(\Gamma_s)}{\log 2s}$$

The entropies of the infinite curve Γ are defined by letting s go to infinity:

$$h^*(\Gamma) = \limsup_{s \to \infty} \frac{S(\Gamma_s)}{\log 2s}$$

$$h_*(\Gamma) = \liminf_{s \to \infty} \frac{S(\Gamma_s)}{\log 2s}$$

and

$$H^*(\Gamma) = \limsup_{s \to \infty} \frac{S_t(\Gamma_s)}{\log 2s}$$

$$H_*(\Gamma) = \liminf_{s \to \infty} \frac{S_t(\Gamma_s)}{\log 2s}.$$

Our theorem 7 shows that

$$0 \le h_*(\Gamma) \le h^*(\Gamma) \le H^*(\Gamma) \le 1.$$

A low entropy curve $(h^*(\Gamma) = 0)$ can be thought of as "deterministic". A high entropy curve $(h^*(\Gamma) = 1)$ is to be identified with a chaotic phenomenon. An infinite straight line, the spiral $\rho = \exp\theta$ or any curve which goes "rapidly" to infinity is deterministic. A very "wiggly" curve, the spiral $\rho = \log\theta$, or any curve which "meanders" will have entropy 1. The entropy of $\rho = \theta^\alpha$ is $(1+\alpha)^{-1}$.

Theorem 8. *Dragon curves have entropy equal to 1/2 ($H_* = H^* = 1/2$). Self avoiding curves drawn on the lattice* $\mathbf{Z}^2$ *have entropy* H^* *at most 1/2.*

We will not prove this result which is really the same as Theorem 4. Dragon curves are halfway in between determinism and chaos. Among self avoiding curves on $\mathbf{Z}^2$, they are however as chaotic as can be.

§9. Entropy and dimension

As Mandelbrot makes it clear in his book [6], dimension measures the complexity. We wish to make this statement precise by showing that dimension and entropy are indeed linked, at least in the context of planar curves.

Recall that $N_s(\delta)$ represents the average number of intersection points of Γ_s with random circles of radius δ. The behaviour of $N_s(\delta)$ is controlled by the number

$$\gamma = \lim_{\delta \to 0} \limsup_{s \to \infty} \frac{\log N_s(\delta)}{\log \operatorname{diam} \Gamma_s}.$$

If $\gamma > 0$, then Γ is not resolvable.

Theorem 9.

$$h_*(\Gamma) \leq 1 - \frac{1}{\operatorname{dim}_* + \gamma}.$$

Proof. We first observe that

$$2 \operatorname{diam} \Gamma_s \leq |\partial K_s| \leq \pi \operatorname{diam} \Gamma_s,$$

so that as s goes to infinity

$$\log \operatorname{diam} \Gamma_s \sim \log |\partial K_s|.$$

Then

$$\operatorname{dim}_* \Gamma = \lim_{\delta \to 0} \liminf_{s \to \infty} \frac{\log \operatorname{area} \Gamma_s(\delta)}{\log |\partial K_s|}$$

$$\geq \liminf_{s \to \infty} \frac{\log s}{\log |\partial K_s|} - \lim_{\delta \to 0} \limsup_{s \to \infty} \frac{\log N_s(\delta)}{\log |\partial K_s|},$$

the last term on the right hand side is γ. As for the first term, it can be easily expressed in terms of $H_*(\Gamma)$. Indeed,

$$H_*(\Gamma) = \liminf_{s \to \infty} \frac{\log 2s/|\partial K_s|}{\log 2s}$$

$$= 1 - \limsup_{s \to \infty} \frac{\log |\partial K_s|}{\log s}$$

hence

$$\limsup_{s \to \infty} \frac{\log |\partial K_s|}{\log s} = 1 - H_*(\Gamma)$$

$$\liminf_{s \to \infty} \frac{\log s}{\log |\partial K_s|} = \frac{1}{1 - H_*(\Gamma)}$$

Finally

$$\dim_*(\Gamma) \geq \frac{1}{1 - H_*(\Gamma)} - \gamma$$

$$H_*(\Gamma) \leq 1 - \frac{1}{\dim_* \Gamma + \gamma}.$$

The proof is completed by noticing that

$$h_*(\Gamma) \leq H_*(\Gamma). \qquad \text{Q.E.D.}$$

Corollary. *If* $h_*(\Gamma) > 1/2$ *, then the curve* Γ *is not resolvable.*

Proof. Indeed, if $h_*(\Gamma) > 1/2$ then

$$\dim_* \Gamma + \gamma > 2$$

hence

$$\gamma > 2 - \dim_* \Gamma \geq 0 \qquad \text{Q.E.D.}$$

A curve such that $h_*(\Gamma) > 1/2$ is somehow close to chaos. A nonresolvable is a curve which has a tendency to "wiggle" and squeeze onto itself. It may be difficult to decide on which branch of Γ one is. One is confused. Our corollary thus establishes that chaos implies confusion.

§10. References

[1] C. Davis, D. Knuth, 'Number representations and dragon curves I,II'. *Jour. Recreational Math.* **3**(1970) 61–81, 113–149.

[2] F.M. Dekking, M. Mendes France, 'Uniform distribution modulo one: a geometrical viewpoint', *J. Reine Angew. Math.* **329** (1981) 143–153.

[3] M. Dekking, M. Mendes France, A. van der Poorten, 'Folds!, I,II,III,' *Mathemat. Intellig.*, 4 (1982) 130–138, 173–181, 190–195.

[4] Y. Dupain, T. Kamae, M. Mendes France, 'Can one measure the temperature of a curve?', *Archive Rat. Mech. and Anal.*, **94** (1986) 155–163 (Corrigenda 98, 1987, 395).

[5] J. Loxton, A. van der Poorten, 'Arithmetic properties of the solution of a class of functional equations', *J. Reine Angew. Math.* **330** (1982) 159–172.

[6] B. Mandelbrot, *Les objets fractales*, Flammarion 1984.

[7] M. Mendes France, 'Folding Paper and Thermodynamics', *Physics Reports, Review section of Physics letters* **103** (1984) 161–172.

[8] M. Mendes France, Dimension and Entropy of Regular Curves. *Publication of RIMS* , Kyoto, (1987).

[9] M. Mendes France, A. van der Poorten, 'Arithmetic and analytic properties of paperfolding sequences', *Bull. Australian Math. Soc.* **24** (1981) 123–130.

[10] M. Mendes France, G. Tenenbaum, 'Dimension des courbes planes, papiers pliés, suites de Rudin-Shapiro'. *Bull. Soc. Math. France* **109** (1981) 207–215.

[11] L. Santaló, *Integral geometry and geometric probability*, Encyclopedia of Mathematics, Addison Wesley 1976.

[12] H. Steinhaus, 'Length, shape and area', *Colloquium Mathem.*, **3** (1954) 1–13.

10
The Riemann hypothesis and the Hamiltonian of a quantum mechanical system

J. V. Armitage
University of Durham, Durham, UK

§1. Introduction

The basic theme of this lecture[1] is an approach to the Riemann Hypothesis in terms of diffusion processes, which has occupied the author's attention for twelve years and which, if correct, has some tantalisingly appealing features culminating in a plausible conjecture that implies the truth of that most celebrated of hypotheses. The connection with diffusion processes suggests that a change of variables (the introduction of imaginary time) might yield a connection with quantum mechanics. That variation offers a possible answer to a conjecture of Berry [1] relating the zeros of the Riemann zeta-function to the Hamiltonian of some quantum mechanical system, which, in its turn makes precise Hilbert's original suggestion that the zeros are eigen-values of some operator and the Riemann Hypothesis is true because that operator is Hermitian. We shall offer possible candidates for Hilbert's operator and Berry's Hamiltonian, but we do not claim satisfactorily to have settled those questions, let alone to have proved the Riemann Hypothesis.

---

[1] The account given here is a slightly extended version of the lecture as actually presented. The author intends to publish a detailed and greatly expanded version elsewhere and hopes that that will not be long delayed. It is worth noting that the approach described here is applicable also to other zeta functions.

§2. The Riemann zeta-function $\zeta(s)$ and the resolvent formula of a semi-group

We begin with Riemann's formula (see [7], pp. 21 and 22)

$$Z(s) = 2s(s-1)\pi^{-s/2}\Gamma(s/2)\zeta(s) \tag{2.1}$$

$$= 2 + 2s(s-1)\int_1^\infty (t^{-\frac{1}{2}-\frac{1}{2}s} + t^{\frac{1}{2}s-1})\psi(t)\,dt$$

where

$$2\psi(t) = \vartheta(t) - 1 = 2\sum_{n=1}^\infty e^{-\pi n^2 t}. \tag{2.2}$$

We recall that (2.1) depends upon the fundamental Jacobi transformation formula

$$1 + 2\psi(t) = t^{-\frac{1}{2}}\left(1 + 2\psi(t^{-1})\right) \tag{2.3}$$

of the theta-function $\vartheta(t)$ (cf. [7], loc.cit.).

The formula (2.1) yields the functional equation $Z(s) = Z(1-s)$, that is, symmetry in the line $\operatorname{Re} s = \frac{1}{2}$. The hypothesis is that the non-trivial zeros of $Z(s)$ (those belonging to $\zeta(s)$) all lie on the line $\operatorname{Re} s = \frac{1}{2}$; that is, are of the form $s = \frac{1}{2} + it$, $t \in \mathbf{R}$.

The fundamental equation (2.3) may be obtained from the fundamental solution

$$u(x,t) = \vartheta(x,t) = \sum_{n=-\infty}^\infty e^{-\pi n^2 t} e^{2\pi inx} \tag{2.4}$$

of the heat (diffusion) equation

$$\frac{1}{4\pi}\frac{\partial^2 u}{\partial x^2} = \frac{\partial u}{\partial t} \qquad u(x,0) = \dot{\delta}(0), \tag{2.5}$$

on the unit circle S^1 (where $\dot{\delta}$ denotes the periodic Dirac delta distrbution), by comparing it with the solution of the same problem on the universal cover of S^1, namely on the real line.

On writing $t = \exp T$ in (2.1) we obtain

$$Z(s) = 2 + 2s(s-1) \int_0^\infty \left\{ \exp\left(\frac{1-s}{2}T\right) + \exp\left(\frac{s}{2}T\right) \right\} \psi(\exp T)\, dT.$$

$$\tag{2.6}$$

That substitution suggests that we write

$$t = \exp T, \qquad x = X \exp(T/2)$$

in the diffusion equation (2.5) which then becomes

$$\frac{1}{4\pi}\frac{\partial^2}{\partial X^2}U(X,T) + \frac{1}{2}X\frac{\partial}{\partial X}U(X,T) = \frac{\partial}{\partial T}U(X,T), \tag{2.7}$$

where $U(X,T) = u(X\exp(\frac{1}{2}T), \exp T)$. The initial condition is $U(X,0) = \vartheta(X,1)$ and at time T the solution is periodic in X with period $\exp(-\frac{1}{2}T)$ — the observer is retreating from the process so that it appears to be contracting exponentially.

The equation (2.7) is closely related to the equation

$$\frac{1}{4\pi}\frac{\partial^2 U}{\partial X^2} + \frac{1}{2}\frac{\partial}{\partial X}(XU) = \frac{\partial U}{\partial T}, \tag{2.8}$$

which is a special case of the Fokker-Planck equation and which was first studied by Smoluchowski in connection with the Brownian motion of a particle subject to an external force. In its discrete form (see §5 below), it was studied by Ornstein and Uhlenbeck in describing the velocity of a particle in Brownian motion. Schrödinger and Kohlrausch pointed out the connection between the Ehrenfest model in thermodynamics and the Brownian motion of an elastically bound particle. (For further details, see the expository article by Kac, [4].)

For our present purposes (mainly to acccommodate the notation used in (2.2)) we write (2.7) as

$$\frac{1}{4\pi}\frac{\partial^2 V}{\partial X^2} + \frac{1}{2}X\frac{\partial V}{\partial X} = \frac{\partial V}{\partial T}, \quad V(X,T) = U(X,T) - 1. \tag{2.9}$$

Now, from (2.6), we have

$$Z(s) - 2 = 2s(s-1) \int_0^\infty \left\{ \exp\left(\frac{1-s}{2}T\right) + \exp\left(\frac{s}{2}T\right) \right\} V(0,T)\, dT,$$

$$\tag{2.10}$$

which is the sum of two Laplace transforms.

We introduce the notation

$$2\Lambda = \frac{1}{2\pi}\frac{\partial^2}{\partial X^2} + X\frac{\partial}{\partial X}$$

and we write (2.9) as

$$\Lambda V(X,T) = \frac{\partial}{\partial T}V(X,T).$$

This last equation defines a contraction semi-group and the resolvent formula applied to it yields:

$$(Z(s)-2)/(s(s-1)) = \left\{ R(-\frac{s}{2},\Lambda) + R(-\frac{1-s}{2},\Lambda)\right\} V(X,0)|_{X=0}$$

(2.11)

where $R(\lambda,\Lambda)$ is to be thought of as $(\lambda - \Lambda)^{-1}$, $V(X,0)$ is the initial distribution and the notation means that the expression on the right hand side is to be evaluated at $X = 0$. (The preceding argument is no more than formal, but it can be justified in terms of a more detailed analysis analogous to that used to define $Z(s)$ in the critical strip.)

The sum of the resolvents in (2.11) may be expressed as a single resolvent, as follows. Write $s = \frac{1}{2} + i\tau$, $\tau \in \mathbb{C}$. Then

$$\frac{Z(\frac{1}{2}+i\tau)-2}{1+4\tau^2} = R(-\tau^2,(2\Lambda+\frac{1}{2})^2)(2\Lambda+\frac{1}{2})V(X,0)|_{X=0}. \quad (2.12)$$

The formulae (2.11) and (2.12) suggest two possible approaches to the Riemann Hypothesis, each of which has interpretations in terms of dynamical systems and both perhaps have connections with the ideas of Berry [1], to which allusion already has been made.

§3. The zeros of $\zeta(s)$ as eigen-values of a family of operators

Suppose that $s = \frac{1}{2} + i\tau$, $\tau \in \mathbf{C}$ is a non-trivial zero of $Z(s)$. The expression (2.11) of $Z(\frac{1}{2} + i\tau)/(1 + 4\tau^2)$ in terms of resolvents suggests the introduction of a function

$$\Phi(X,\tau) = \frac{Z(X,\frac{1}{2} + i\tau) - 2(1 - \exp(-\pi X^2))}{1 + 4\tau^2} \tag{3.1}$$

$$= -4\left\{R(-\tfrac{1}{4} - \tfrac{1}{2}i\tau,\Lambda) + R(-\tfrac{1}{4} + \tfrac{1}{2}i\tau,\Lambda)\right\} V(X,0)$$
$$+ 2\exp(-\pi X^2)/(1 + 4\tau^2)$$

$$= 2\int_0^\infty \left((\exp(\tfrac{1}{4} + \tfrac{1}{2}i\tau)T + \exp(\tfrac{1}{4} - \tfrac{1}{2}i\tau)T)\right.$$

$$\left.\times \sum_{n=1}^\infty \exp(-\pi n^2 e^T)\cos(2\pi n e^{\frac{1}{2}T}X)\right)dT$$

$$+ \frac{2\exp(-\pi X^2)}{1 + 4\tau^2}$$

and of a function

$$\gamma(X) = \tfrac{1}{2}\exp(-\pi X^2) + (2\Lambda + \tfrac{1}{2})\sum_{n\neq 0}\exp(-\pi n^2)\exp(2\pi i nX);$$
$$\tag{3.2}$$

so that (see (2.12) above):

$$\left((2\Lambda + \tfrac{1}{2})^2 + \tau^2\right)\Phi(X,t) = \gamma(X). \tag{3.3}$$

Moreover, since $\frac{1}{2} + i\tau$ is a zero of $Z(s)$

$$\Phi(0,\tau) = 0 \tag{3.4}$$

and it is readily verified that

$$\Phi^{(1)}(0,\tau) = \Phi^{(3)}(0,\tau) = 0, \quad \lim_{X\to\infty}\left|X\Phi^{(i)}(X,\tau)\right| = O(1),$$

$$0 \le i \le 4. \tag{3.5}$$

Indeed the first two follow from the fact that $\Phi(X, \tau) = \Phi(-X, \tau)$, whilst the fact that $\Phi(X, \tau) = \Phi(X, -\tau)$ generalises the functional equation of $Z(\frac{1}{2} + i\tau)$. Finally we note that

$$\gamma(0) = 0 \qquad \text{and} \qquad \frac{1}{4\pi^2}\Phi^{(4)}(0, \tau) + \frac{3}{2\pi}\Phi^{(2)}(0, \tau) = 0. \quad (3.6)$$

The equation (3.3) and the boundary conditions in (3.4) to (3.6) suggest the application of functional analysis, and in particular of perturbation theory, to the characterisation of the numbers τ satisfying (3.3) and (3.4) to (3.6), which yield the zeros $\frac{1}{2} + i\tau$ of $\zeta(s)$. Obviously one would like to prove that all those τ are real. One possible line of attack (though not necessarily the best) is as follows.

Suppose one could construct some Hilbert space $\mathcal{H}_0$, of functions defined on the internal $I = [0, \infty)$, with inner product $(f, g)_0$, containing $\Phi(X, \tau)$ and $\gamma(X)$. Given $\Phi(X, \tau)$ and $\gamma(X)$ defined by (3.1) and (3.2) suppose that τ corresponds to a zero $\frac{1}{2} + i\tau$ of $\zeta(\frac{1}{2} + i\tau)$. We introduce real valued functions $\alpha(X)$ and $\beta(X)$ such that

$$\Phi(X, \tau) = \alpha(X) + i\beta(X), \qquad \Phi(X, \bar{\tau}) = \alpha(X) - i\beta(X)$$

and a function $\theta(X)$, not identically zero, such that

$$(\beta(X), \theta(X))_0 = 0, \qquad (\alpha(X), \theta(X))_0 \neq 0.$$

With the foregoing notation we define a projection[2] P_τ (depending on τ) of $\mathcal{H}_0$ on the sub-space spanned by $\gamma(X)$ by

$$P_\tau f = -\frac{(f, \theta)_0}{(\alpha, \theta)_0}\gamma, \qquad f \in \mathcal{H}_0. \quad (3.7)$$

It is important to note the dependence of P on τ: there is one projection operator P_τ for each τ such that $\zeta(\frac{1}{2} + i\tau) = 0$. Note also that it follows from (3.7) that

$$P_{\bar{\tau}}f = P_\tau f, \qquad f \in \mathcal{H}_0. \quad (3.8)$$

---

[2] Since $P_\tau^2 \neq P_\tau$ this is not strictly a projection. That could be achieved by introducing a normalising factor, but the fact is not used.

From (3.3) and (3.7) we see that:

$$\left\{ \left(2\Lambda + \tfrac{1}{2} \right)^2 + P_\tau + \tau^2 \right\} \Phi(X, \tau) = 0; \qquad (3.9)$$

so that P_τ defines a perturbation of the differential operator $(2\Lambda + \tfrac{1}{2})^2$ on the space $\mathcal{H}_0$.

A possible strategy for a proof of the Riemann Hypothesis may now be outlined. Given a τ such that $\zeta(\tfrac{1}{2} + i\tau) = 0$, define $\Phi(X, \tau)$, $\gamma(X)$ and P_τ by (3.1), (3.2) and (3.7) Then, by (3.9), $-\tau^2$ is an eigenvalue of the perturbed operator $(2\Lambda + \tfrac{1}{2})^2 + P_\tau$. We would like to show that its eigen-values are real and negative; From which it would follow that τ is real.

It is unfortunate that the perturbation depends on τ ; so that we are working not with a single operator, but with an infinite family of operators. As will emerge in §5 below it is possible to consider a single operator and a perturbation that does not depend on τ , but then the space $\mathcal{H}_0$ becomes problematical. We will return to that theme later in §6, as well as in §5. See also the remarks at the end of §4.1, concerning the possibility of replacing the fourth order operator in (3.3) by a second order stochastic differential operator.

§4. The Fokker-Planck equation and the Hamiltonian of a quantum mechanical system

Montgomery [5] considered the pair-correlation of zeros $\tfrac{1}{2} + it$ of $\zeta(s)$, that is the distribution of $t - t'$ for pairs of zeros. On the basis of the Riemann Hypothesis, he conjectured that the pair-correlation is given by $1 - \left(\sin(\pi u)/\pi u \right)^2$ and Dyson pointed out that the eigen-values of random, complex Hermitian matrices have the same statistics (see [5]). These observations led Berry [1] to investigate complex Hermitian matrices whose elements have a Gaussian distribution invariant under unitary transformations. He showed that the numerical evidence revealed a close connection between their eigen-values, E_j (that is, certain energy levels) and zeros $\tfrac{1}{2} + iE_j$ of $\zeta(s)$. In particular there is a striking similarity between their distribution and that given by the Riemann-Siegel formula.

The foregoing led Berry to conjecture that the E_j are the eigen-values of some Hermitian operator $\hat{H}$, that is the Hamiltonian of a quantum mechanical system such that: (i) $\hat{H}$ has a classical limit; (ii) the classical orbits are all chaotic (unstable); (iii) the classical orbits do not possess time-reversal symmetry,

Here are two recipes for possible candidates for $\hat{H}$.

4.1 The Fokker-Planck Equation.

Consider the basic differential equation (3.3). We note that all the functions involved define tempered distributions on the real line and so we may take Fourier transforms:

$$\hat{f}(Y) = \mathcal{F}(f(X)) = \int\limits_{-\infty}^{\infty} e^{-2\pi i Y X} f(X)\, dX. \qquad (4.1)$$

We obtain the differential equation

$$Y^2 \frac{d^2\hat{\Phi}}{dY^2} + (4\pi Y^3 + 2Y)\frac{d\hat{\Phi}}{dY} + (4\pi^2 Y^4 + \pi Y^2 + \tfrac{1}{4} + \tau^2)\hat{\Phi} = \hat{\gamma}(Y), \qquad (4.2)$$

where $\hat{\Phi}$, $\hat{\gamma}$ are defined by (4.1), (3.1) and (3.2) and in particular

$$\hat{\gamma}(Y) = \tfrac{1}{2}e^{-\pi Y^2} - \sum_{n \neq 0} e^{-\pi n^2} \left[(2\pi Y^2 + 1)\delta_{Y-n} + Y\delta^{(1)}_{Y-n} \right].$$

The equation (4.2) is of the form

$$\frac{1}{2}\frac{\partial^2}{\partial Y^2}\left\{ B_2(Y)\hat{\Phi}(Y) \right\} - \frac{\partial}{\partial Y}\left\{ B_1(Y)\hat{\Phi}(Y) \right\} - $$
$$B_0(Y)\hat{\Phi}(Y) + \tau^2\hat{\Phi}(Y) = \hat{\gamma}(Y),$$

for suitably defined B_0, B_1, B_2 and the left hand side is evidently of the form of the Fokker-Planck equation.

Now if a kernel (propagator) $K(Y, \xi; T)$ satisfies an equation

$$\frac{\partial K}{\partial T} = -B_0(Y)K - \frac{\partial}{\partial Y}[B_1(Y)K] + \frac{1}{2}\frac{\partial^2}{\partial Y^2}[B_2(Y)K], \qquad (4.3)$$

and if the operator U_T:

$$U_T f(Y) = \int\limits_{-\infty}^{\infty} K(Y, \xi; T)f(\xi)\, d\xi$$

defines a one-parameter semi-group and if that is a unitary operator, then its self-adjoint generator $\hat{H}$ satisfies

$$ i\frac{\partial K}{\partial T} = \hat{H}K $$

and is the Hamiltonian. Unfortunately the semi-group here is not unitary but the analogy is tempting. One would need to explore the significance of the 'forcing term' $\hat{\gamma}(Y)$.

There is of course a connection between the Fokker-Planck equation and a first order stochastic differential equation (cf. the paper by Pearle, pp.84—108, in [6] and the excellent survey article by Williams, [8]). It is tempting to suggest that the fourth order equation (3.3) should correspond in the Itô model to a second order stochastic differential equation (a suggestion I owe to Professor Williams).

4.2 The harmonic oscillator.

(The ideas in this section were suggested to the author during the conference by Professor Berry and his student Jonathan Keating.)

We return again to (3.3) and this time we write

$$ \Phi(X,\tau) = e^{-\pi X^2/2}\Psi(X,\tau) $$

(which defines Ψ); so that the equation (3.3) becomes:

$$ \left\{\left(\frac{1}{2\pi}\frac{d^2}{dX^2} - \tfrac{1}{2}\pi X^2\right)^2 + \tau^2\right\}\Psi(X,\tau) = e^{\frac{1}{2}\pi X^2}\gamma(X) \qquad (4.4) $$

or

$$ e^{-\frac{1}{2}\pi X^2}\left\{\left(\frac{1}{2\pi}\frac{d^2}{dX^2} - \tfrac{1}{2}\pi X^2\right)^2 + \tau^2\right\}e^{\frac{1}{2}\pi X^2}\Phi(X,\tau) = \gamma(X). $$

The corresponding homogeneous equation, with $\gamma(X)$ replaced by 0, is Weber's equation. After the substitution $Y = e^{-\pi i/4}X$ the equation (4.4) becomes

$$ \left\{\left(-\frac{i}{2\pi}\frac{d^2}{dY^2} - \tfrac{1}{2}i\pi Y^2\right)^2 + \tau^2\right\}e^{-\frac{1}{2}\pi iY^2}\Phi\left(e^{\pi i/4}Y,\tau\right) $$

$$ = e^{-\frac{1}{2}\pi iY^2}\gamma\left(e^{\pi i/4}Y\right); $$

that is,

$$\left\{ \left(\frac{1}{2\pi}\frac{d^2}{dY^2} + \tfrac{1}{2}\pi Y^2 \right)^2 - \tau^2 \right\} e^{-\frac{1}{2}\pi i Y^2} \Phi\left(e^{\pi i/4} Y, \tau \right)$$

$$= -e^{-\frac{1}{2}\pi i Y^2} \gamma\left(e^{\pi i/4} Y \right);$$

If we now introduce the momentum operator $p = -i\hbar(d/dY)$ and the position operator $q = Y$, we obtain

$$\left\{ \left(-\frac{p^2}{2m} + \tfrac{1}{2}\pi q^2 \right)^2 - \tau^2 \right\} e^{-\frac{1}{2}\pi i Y^2} \Phi\left(e^{\pi i/4} Y, \tau \right)$$

$$= -e^{-\frac{1}{2}\pi i Y^2} \gamma\left(e^{\pi i/4} Y \right), \qquad (4.5)$$

where $2\pi\hbar^2 = 2m$. Although the functions are no longer in $L^2(I)$ (nor are they in the space $\mathcal{H}_0$ introduced in §6 below), the operator on the lefthand side of (4.5) at least formally satisfies Berry's conditions.

§5. A random walk approximation to the Riemann Hypothesis

The connection between random walks and Brownian motion is well known and so also the connection with the Schrödinger equation, on replacing 'time' by 'imaginary time'. In this section we use a random walk approach to the Ornstein-Uhlenbeck process (or the Fokker-Planck equation) to exhibit a polynomial whose zeros, under a suitable limiting process, ought to be the zeros of the Riemann zeta-function. The was indeed the author's original approach to the problem and in some respects it remains the most appealing one.[3]

We follow Kac [4], especially §4, slightly modified to yield (2.7) rather than (2.8). For brevity we shall omit the interpretation in terms of transition probabilities, but that is readily restored. We shall also omit any justification of the crucial 'invarience principles' underlying the limiting processes involved; they lie very deep and can be found in the exposition [8] by Williams.

---

[3] In a letter to Professor Bombieri in 1975, who suggested some improvements incorporated in this account

We consider $2R$ intervals on the real line, each of length $\Delta > 0$, and points $k\Delta$, $-R \le k \le R$, which determine those intervals. We consider the values of a function $V(X, T)$ at the discrete set of points $(k\Delta, s\sigma)$, where $\sigma > 0$, k is an integer satisfying $-R \le k \le R$ and s is a positive integer. The values of the function $V(k\Delta, s\sigma)$ for a given s are given by a vector

$$V(s) = (\ldots, V(k\Delta, s\sigma), \ldots), \qquad -R \le k \le R. \qquad (5.1)$$

We suppose that the components of $V(s)$ and $V(s+1)$ satisfy the difference equation

$$V(k\Delta, (s+1)\sigma) = \tfrac{1}{2}\left(1 + \frac{k}{R}\right) V((k+1)\Delta, s\sigma)$$

$$+ \tfrac{1}{2}\left(1 - \frac{k}{R}\right) V((k-1)\Delta, s\sigma). \qquad (5.2)$$

Hence

$$\frac{V(k\Delta, (s+1)\sigma) - V(k\Delta, s\sigma)}{\sigma}$$

$$= \frac{\Delta^2}{\sigma}\left(\frac{\tfrac{1}{2}\left(1 + \frac{k}{R}\right) V((k+1)\Delta, s\sigma) - V(k\Delta, s\sigma)}{\Delta^2}\right.$$

$$\left. + \frac{\tfrac{1}{2}\left(1 - \frac{k}{R}\right) V((k-1)\Delta, s\sigma)}{\Delta^2}\right). \qquad (5.3)$$

In the limit as

$$\left.\begin{array}{c} \Delta \to 0, \quad \sigma \to 0, \quad R \to \infty, \quad \dfrac{\Delta^2}{2\sigma} = \dfrac{1}{4\pi}, \\[2mm] \dfrac{1}{R\sigma} \to \dfrac{1}{2}, \quad s\sigma \to T, \quad k\Delta \to X, \end{array}\right\} \qquad (5.4)$$

the difference equation (5.3) goes over to the differential equation (cf. (2.9))

$$\frac{\partial V}{\partial T} = \frac{1}{4\pi}\frac{\partial^2 V}{\partial X^2} + \frac{1}{2}X\frac{\partial V}{\partial X} \qquad (5.5)$$

with the initial condition $V(X, 0)$ derived from a vector θ^* whose k^{th} component is the value of $\vartheta(X, 1) - 1$ at the point $X = k\Delta$. We note in passing that θ^* itself may be obtained from a random walk approximation to Brownian motion, using the De Moivre-Laplace limit theorem (see [4],§2).

Let C be the $(2R+1) \times (2R+1)$ matrix:

$$C = \begin{pmatrix} 0 & 0 & 0 & 0 & 0 & \cdots & 1 \\ 1 - \frac{1}{2R} & 0 & \frac{1}{2R} & 0 & 0 & \cdots & 0 \\ 0 & 1 - \frac{2}{2R} & 0 & \frac{2}{2R} & 0 & \cdots & 0 \\ 0 & 0 & 1 - \frac{3}{2R} & 0 & \frac{2}{2R} & \cdots & 0 \\ \vdots & \vdots & \vdots & \vdots & \vdots & \ddots & \vdots \\ 1 & 0 & 0 & 0 & 0 & \cdots & 0 \end{pmatrix}.$$

$$(5.6)$$

We observe that C is in the Markoff algebra and that

$$V(s+1) = CV(s), \quad V(s) = C^s V(0) \tag{5.7}$$

So the matrix C^s affords the discrete analogue of the semi-group defined by (5.5). The corresponding 'infinitesimal generator' is the matrix

$$L = \frac{R}{2}(C - I) = \frac{1}{2\pi\Delta^2}(C - I). \tag{5.8}$$

In place of the resolvent $R(\lambda, \Lambda)$ we have the matrix $(\lambda - L)^{-1}$ and so in place of the right hand side of (2.11), with $s = \frac{1}{2} + i\tau$, $\tau \in \mathbf{C}$, we have

$$-2\left\{(i\tau I + \tfrac{1}{2}I + 2L)^{-1} + (-i\tau I + \tfrac{1}{2}I + 2L)^{-1}\right\}\theta^* \tag{5.9}$$

and we require the element in the middle row of the column $(2R+1)$-vector, namely the element corresponding to $X = k\Delta$, with $k = 0$. Comparison with (2.11) suggests that we should define $Z(\frac{1}{2} + i\tau)$ by

$$Z(\tfrac{1}{2} + i\tau) - 2 = 2(\tfrac{1}{4} + \tau^2)\Big\{ (2L + (\tfrac{1}{2} + i\tau)I)^{-1}$$

$$+ (2L + (\tfrac{1}{2} - i\tau)I)^{-1} \Big\}\theta^*\Big|_{k=0} \tag{5.10}$$

where the notation means that we take the element corresponding to $k = 0$ in the vector on the right hand side of (5.10). We observe that

$$\det(zI + 2L) = z(z-1)(z-2)\ldots(z-2R)$$

so the inversion of the matrices is justified provided that $\frac{1}{2} \pm i\tau \notin Z$ (and those excluded values include the trivial zeros of $Z(\frac{1}{2} + i\tau)$).

Let ξ, η be vectors in $\mathbf{C}^{2R+1}$ such that:

$$\left.\begin{aligned}(2L + (\tfrac{1}{2} + i\tau)I)\,\xi &= 2(\tfrac{1}{4} + \tau^2)\theta^*, \\ (2L + (\tfrac{1}{2} - i\tau)I)\,\eta &= 2(\tfrac{1}{4} + \tau^2)\theta^*. \end{aligned}\right\} \tag{5.12}$$

If ξ has components $x_{-R}, \ldots, x_R$ and η has components $y_{-R}, \ldots, y_R$, it follows from (5.10) that, if $Z(\frac{1}{2} + i\tau) = 0$ and if $\frac{1}{2} \pm i\tau \notin \mathbf{Z}$, then we require those values of τ for which

$$x_0 + y_0 = -2. \tag{5.13}$$

Assuming that under the limiting process (5.3) the zeros of $Z(\frac{1}{2} + i\tau)$ as defined by (5.10) tend to the zeros of Riemann's function $Z(\frac{1}{2} + i\tau)$ as defined by (2.1), we can say that in order to prove the Riemann Hypothesis it suffices to prove that the numbers τ defined by (5.12) and (5.13) are real.

There are several ways of proceeding from that observation. For example, define the polynomials $P(z)$, $Q(z)$ by

$$\left.\begin{aligned}Q(z) &= \det(zI + \tfrac{1}{2}I + 2L) \\ P(z) &= \det(zI + \tfrac{1}{2}I + 2L)_\theta \end{aligned}\right\} \tag{5.14}$$

where $(zI + \frac{1}{2}I + 2L)_\theta$ denotes that in the matrix $(zI + \frac{1}{2}I + 2L)$ the middle column is replaced by $(\frac{1}{2} - 2z^2)\theta^*$. Let

$$F(z) = P(-z)Q(z) + Q(-z)P(z) + 2Q(-z)Q(z). \tag{5.15}$$

Then it follows from (5.14), (5.12) and (5.13), that we require those values of $z = i\tau$ for which $F(z) = F(i\tau) = 0$. We note that $F(z)$ is a polynomial in z^2 of degree $2R + 1$; so it suffices to prove that that polynomial in z^2 has *negative* roots, that is, that $(i\tau)^2 = z^2 < 0$, whence $\tau^2 > 0$.

We have already noticed that the sum in (5.10) may be replaced by a single resolvent and we observe that, if ξ and η are defined by (5.12), then

$$\xi + \eta = \left(\tau^2 I + (2L + \tfrac{1}{2}I)^2\right)^{-1}(2L + \tfrac{1}{2}I)(1 + 4\tau^2)\,\theta^*. \tag{5.16}$$

As before we require $x_0 + y_0 = -2$. Now by Cramer's rule,

$$x_0 + y_0 = \frac{\det\left(\tau^2 I + \left(2L + \frac{1}{2}I\right)^2\right)_\theta}{\det\left(\tau^2 I + \left(2L + \frac{1}{2}I\right)^2\right)}$$

where now in the numerator the middle column of the matrix $\tau^2 I + \left(2L + \frac{1}{2}I\right)^2$ is replaced by $\left(1 + 4\tau^2\right)\left(2L + \frac{1}{2}I\right)\theta^*$. So, if τ corresponds to a zero of $\zeta(\frac{1}{2} + i\tau)$ and therefore $x_0 + y_0 = -2$, we require

$$\det A(\tau^2) = 0, \tag{5.18}$$

where $A(\tau^2)$ is the matrix $\left(2L + \frac{1}{2}I\right)^2 + \tau^2 I$ with the vector $\left(\frac{1}{2} + 2\tau^2\right)\left(2L + \frac{1}{2}I\right)\theta^*$ added to the middle column. On noting that the middle component of the vector $\left(2L + \frac{1}{2}I\right)\theta^*$ is $-1/2$ (cf.(3.6)), from which it follows that the middle element of the matrix $A(\tau^2)$ is independent of τ^2, one can now use elementary row and column operations on (5.18) to obtain a matrix whose eigen-values are the values of $-\tau^2$ corresponding (by means of the limiting process (5.4)) to the zeros of $\zeta(\frac{1}{2} + i\tau)$.

§6.   A tentative application of perturbation theory to suggest a possible proof of the Riemann Hypothesis

We return to the formulation of the Riemann Hypothesis given in §3 above. We consider the differential operator

$$\left(\frac{1}{2\pi}\frac{d^2}{dX^2} + X\frac{d}{dX} + \frac{1}{2}\right) \tag{6.1}$$

and its formal adjoint

$$\left(\frac{1}{2\pi}\frac{d^2}{dX^2} - X\frac{d}{dX} + \frac{1}{2}\right) \tag{6.2}$$

We consider the boundary problem in $L_2(I)$ defined by

$$\left\{\left(2\Lambda + \frac{1}{2}\right) + \rho^2\right\}\phi(X,\rho) = 0$$

and the conditions (3.4), (3.5) and (3.6) above. It turns out that a complete set of eigenfunctions is given by:

$$\phi_n(X,\rho_n) = C_n \left\{ \pi^{-\frac{1}{2}i\rho_n} \Gamma(\tfrac{1}{4} + \tfrac{1}{2}i\rho_n) \, {}_1F_1(\tfrac{1}{4} + \tfrac{1}{2}i\rho_n, \tfrac{1}{2}, -\pi X^2) \right.$$
$$\left. + \pi^{\frac{1}{2}i\rho_n} \Gamma(\tfrac{1}{4} - \tfrac{1}{2}i\rho_n) \, {}_1F_1(\tfrac{1}{4} - \tfrac{1}{2}i\rho_n, \tfrac{1}{2}, -\pi X^2) \right\},$$

(6.3)

where ${}_1F_1$ denotes the confluent hypergeometric function and the C_n are normalising factors. The adjoint eigenfunctions are:

$$\phi_n^*(X,\rho_n) = D_n \left\{ \pi^{-\frac{1}{2}i\rho_n} \Gamma(\tfrac{1}{4} + \tfrac{1}{2}i\rho_n)\Gamma(\tfrac{3}{4} + \tfrac{1}{2}i\rho_n) \right.$$
$$\times \psi(\tfrac{1}{4} + \tfrac{1}{2}i\rho_n, \tfrac{1}{2}, \pi X^2)$$
$$+ \pi^{\frac{1}{2}i\rho_n} \Gamma(\tfrac{1}{4} - \tfrac{1}{2}i\rho_n)\Gamma(\tfrac{3}{4} - \tfrac{1}{2}i\rho_n)$$
$$\left. \times \psi(\tfrac{1}{4} - \tfrac{1}{2}i\rho_n, \tfrac{1}{2}, \pi X^2) \right\}, \quad (6.4)$$

where ψ denotes Kummer's function and the D_n are normalising factors. (See [3] for the properties of ${}_1F_1$ and ψ used here.) The numbers ρ_n are real and satisfy

$$\pi^{-\frac{1}{2}i\rho_n} \Gamma(\tfrac{1}{4} + \tfrac{1}{2}i\rho_n) = -\pi^{\frac{1}{2}i\rho_n} \Gamma(\tfrac{1}{4} - \tfrac{1}{2}i\rho_n) \qquad (6.5)$$

(as follows from (3.4) and the fact that $\tfrac{1}{2}+i\tau$ is a zero of $\zeta(\tfrac{1}{2}+i\tau)$). The system defined by (6.3), (6.4) and (6.5) forms a complete bi-orthonormal system, namely, in the usual L_2–norm,

$$(\phi_m, \phi_n^*) = \begin{cases} 1, & m = n, \\ 0, & m \neq n. \end{cases}$$

The functions ϕ_n and ϕ_m^* are however highly correlated, since

$$(\phi_m, \phi_n) \neq 0 \quad \text{and} \quad (\phi_m^*, \phi_n^*) \neq 0.$$

We note in passing a nice generalisation of the approximate functional equation for $\zeta(\tfrac{1}{2} + i\tau)$, which appears to be new. The sketch which follows is purely formal and contains only a recipe for a proof.

We have, from (3.1)

$$(1 + 4\tau^2)\Phi(X,\tau) = 2e^{-\pi X^2} - (1 - 4\tau^2)\int_1^\infty \left\{ t^{-\frac{3}{4}}\left(t^{\frac{1}{2}i\tau} + t^{-\frac{1}{2}i\tau}\right)\right.$$

$$\left.\times \sum_{n=1}^\infty e^{-\pi n^2 t}\cos(2\pi n\sqrt{t}X)\right\} dt \qquad (6.6)$$

Now (see [3], p272, (5))

$$\int_0^\infty e^{-\pi n^2 t}t^{-\frac{3}{4}}t^{\frac{1}{2}i\tau}\cos(2\pi n\sqrt{t}X)\,dt =$$

$$n^{-\frac{1}{2}-i\tau}\pi^{\frac{1}{4}-\frac{1}{2}i\tau}\Gamma(\tfrac{1}{4} + \tfrac{1}{2}i\tau){}_1F_1(\tfrac{1}{4} + \tfrac{1}{2}i\tau, \tfrac{1}{2}, -\pi X^2).$$

So

$$2\int_1^\infty e^{-\pi n^2 t}t^{-\frac{3}{4}}\cos(\tfrac{1}{2}\tau\log t)\cos(2\pi n\sqrt{t}X)\,dt =$$

$$2n^{-\frac{1}{2}-i\tau}\pi^{\frac{1}{4}-\frac{1}{2}i\tau}\Gamma(\tfrac{1}{4} + \tfrac{1}{2}i\tau){}_1F_1(\tfrac{1}{4} + \tfrac{1}{2}i\tau, \tfrac{1}{2}, -\pi X^2)$$

$$+ 2n^{-\frac{1}{2}+i\tau}\pi^{\frac{1}{4}+\frac{1}{2}i\tau}\Gamma(\tfrac{1}{4} - \tfrac{1}{2}i\tau){}_1F_1(\tfrac{1}{4} - \tfrac{1}{2}i\tau, \tfrac{1}{2}, -\pi X^2)$$

$$- 2\int_0^1 e^{-\pi n^2 t}t^{-\frac{3}{4}}\cos(\tfrac{1}{2}\tau\log t)\cos(2\pi n\sqrt{t}X)\,dt. \qquad (6.7)$$

On combining (6.6) and (6.7) formally, one obtains an approximate functional equation for $\Phi(X,\tau)$. Of course the interchange of summation and integration implicit in the foregoing is not valid; to obtain the correct version and proof, one should follow the argument used by Titchmarsh ([7], Chapter IV) to prove his (4.12.4) and Theorem 4.14 (op.cit.pp.79-84).

We return to our main theme!

We refer to the strategy for a proof of the Riemann Hypothesis given in §3 above. We begin by asserting that $\phi_n(X, \rho_n)$ and $\phi_n^*(X, \rho_n)$ defined by (6.3), (6.4) and (6.5) have the property that

$$\phi_n(X, \rho_n) = O\left(X^{-5/2+\epsilon}\right), \quad \phi_n^*(X, \rho_n) = O\left(X^{-5/2+\epsilon}\right), \qquad (6.8)$$

(for $\epsilon > 0$) where the O-symbol involves dependence on ρ_n. It follows from (6.8) and the definition of $\gamma(X)$ in (3.2) that the integrals

$$\int_0^\infty \phi_n \bar{\gamma} \, dX \quad \text{and} \quad \int_0^\infty \phi_n^* \bar{\gamma} \, dX$$

are defined and we may accordingly denote them by (ϕ_n, γ), (ϕ_n^*, γ), respectively.

We are now in a position to define the fundamental space $\mathcal{H}_0$. Let H_{B^2} denote the space of functions $f \in L_2(I)$ such that $f^{(1)}$, $f^{(2)}$, $f^{(3)}$ and $f^{(4)}$ exist and are absolutely continuous in I and moreover $B^2 f = (2\Lambda + \frac{1}{2})^2 f \in L_2(I)$. Let $\hat{H}_{B^2}$ denote the completion of H_{B^2} with respect to the norm $\|f\|_1 = \|f\| + \|B^2 f\|$, where $\|f\|$ denotes the L_2-norm. Finally define,

$$\mathcal{H}_0 = \hat{H}_{B^2} \oplus \{\Phi(X, \tau)\} \oplus \{\Phi(X, \bar{\tau})\} \oplus \{\gamma\}, \tag{6.9}$$

where $\{f\}$ denotes the C-linear space spanned by f and $\oplus$ denotes a direct sum of C-linear spaces. Note that $(B^2 + \tau^2)\Phi(X, \tau) = \gamma(X)$; so that $\Phi(X, \tau) \notin \hat{H}_{B^2}$, since $\gamma(X) = \Omega(X)$. (The notation is that of Titchmarsh, [7], Chapter VIII and denotes that $|\gamma(X)| > AX$ for some arbitrarily large X.) We define an inner product in $\mathcal{H}_0$ by

$$(f, g)_0 = \sum_{n=1}^\infty (f, \phi_n^*)\overline{(g, \phi_n)}. \tag{6.10}$$

For $f \in \hat{H}_{B^2}$, there is an eigen-function expansion

$$f = \sum_n (f, \phi_n^*)\phi_n$$

and so the definition (6.10) is consistent with Parseval's formula when that is applicable. In particular, we have

$$\|\gamma\|_0 = \sum_{n=1}^\infty (\gamma, \phi_n^*)\overline{(\gamma, \phi_n)}, \tag{6.11}$$

the terms of the series on the right behaving like $\rho_n^{-2} \log n$, where $\rho_n \sim n/(\log n)$, for n large enough.

So $\mathcal{H}_0$ may be given the structure of a Hilbert space with inner product given by (6.10).

Now we recall the fundamental abstract perturbation theorem of Schwartz (see [2], Part III, Chapter XIX, Theorem 7).

Let T be a spectral operator, in a Hilbert space, with spectrum λ_n and let d_n denote the distance of λ_n from the remainder of the spectrum. Suppose that the spectral measure E defines a projection $E(\lambda_n)$ onto a 1-dimensional subspace and that $\sum E(\lambda_i) = I$, where the sum is taken over the spectrum and I denotes the identity. Let P be a bounded perturbation. Then:

(a) if $\sum d_n^{-2}$ converges, $T + P$ is spectral;

(b) if C_n is the circle $|z - \lambda_n| = \frac{1}{2}d_n$ and if σ_n denotes that part of the spectrum of $T + P$ lying inside C_n, then if n is large enough, σ_n consists of a single point.

The idea of our 'proof' is to apply the theorem to the perturbation P defined by (3.7) and the operator T defined by $(2\Lambda + \frac{1}{2})^2$ in $\mathcal{H}_0$.

For the unperturbed operator $(2\Lambda + \frac{1}{2})^2$ in $\mathcal{H}_0$ we have the spectrum $\{\lambda_n\}$, given by $\lambda_n = -\rho_n^2$ where $\rho_n \in \mathbf{R}$ satisfies (6.5). We have $\rho_n \sim n/(\log n)$ and so

$$d_n \sim (\rho_{n+1} - \rho_n)(\rho_{n+1} + \rho_n) \sim (2n)/(\log n)^2.$$

Hence $\sum d_n^{-2}$ converges. We take $P = P_\tau = P_{\overline{\tau}}$ where P_τ is defined by (3.7) (cf.(3.8)). The dependence of n on P in (b) suggests that the dependence of P_τ on τ might invalidate the appeal to Schwartz's theorem. However an examination of the proof of that theorem and explicit calculations for the constants involved reveals that it may be applied.

The projection $E(\lambda_n)$ is given in our case by

$$E(\lambda_n)f = E(-\rho_n^2)f = (f, \phi_n^*)\phi_n.$$

So the projection is one-dimensional, but the condition $\sum E(\lambda_i) = I$ requires

$$\sum (f, \phi_n^*)\phi_n = f.$$

Now that is true (by the remark following (6.10) above) for $f \in \hat{H}_{B^2}$ and indeed for f in the spaces spanned by $\Phi(X, \tau)$ or $\Phi(X, \overline{\tau})$

but

$$\sum_{n=1}^{\infty} (\gamma, \phi_n^*)\phi_n \to \gamma \tag{6.12}$$

uniformly only on bounded sets. For the interval $I = [0, \infty)$, (6.12) is true only in the sense of conditional convergence.

Let us suppose however that the theorem of Schwartz is applicable under that weaker hypothesis. Suppose that τ gives a zero $\frac{1}{2} + \frac{1}{2}i\tau$ of the Riemann zeta-function and suppose that $\tau \neq \bar{\tau}$. Then by (3.7) and (3.9)

$$\left\{ \left(2\Lambda + \tfrac{1}{2}\right)^2 + P_\tau + \tau^2 \right\} \Phi(X, \tau) = 0,$$

and

$$\left\{ \left(2\Lambda + \tfrac{1}{2}\right)^2 + P_\tau + \bar{\tau}^2 \right\} \Phi(X, \bar{\tau}) = 0,$$

since $P_{\bar{\tau}} = P_\tau$. If $\tau \neq \bar{\tau}$, we obtain a contradiction to (b) above. It follows that $\tau = \bar{\tau}$ and so all the zeros are real and simple and the Riemann Hypothesis is (or rather, would be!) true.

§7. References

[1] M. V. Berry 'Riemann's zeta function: a model of quantum chaos?' in *Quantum Chaos and Statistical Nuclear Physics*, Lecture Notes in Physics 263 (Springer, Berlin 1986).

[2] N. Dunford and J. T. Schwartz, *Linear Operators*, Parts I, II and III, (Wiley Interscience, New York, 1971).

[3] A. Erdélyi, *Higher Transcendental Functions*, Volume I (McGraw Hill, New York, 1953).

[4] M. Kac, 'Random walk and the theory of Brownian motion', *Americal Mathematical Monthly*, **54** (1947), 369–391.

[5] H. L. Montgomery, 'The pair correlation of zeros of the zeta function', *Analytic number theory* (Proc. Symp. Pure Math. Vol.XXIV), 181–193, (Amer. Math. Soc., Providence, R.I. 1973).

[6] P. Pearle, 'Models for reduction', in *Quantum Concepts in Space and Time*, (ed. R. Penrose and C. J. Isham), 84–108, (Clarendon Press, Oxford, 1986).

[7] E. C. Titchmarsh, *The Theory of the Riemann Zeta-function* (2nd Ed., revised by D. R. Heath-Brown), Clarendon Press, Oxford,1986.

[8] D. Williams, 'Brownian motions and diffusions as Markov processes', *Bull. London Math. Soc.*, **6** (1974) 257–303.

Printed in the United States
By Bookmasters